FIRST EDITION

RELIGION AND CONTEMPORARY ISSUES: POLITICS, ECOLOGY, AND WOMEN'S RIGHTS

Edited by Ivanessa Arostegui

Bassim Hamadeh, CEO and Publisher
Kassie Graves, Director of Acquisitions
Jamie Giganti, Senior Managing Editor
Miguel Macias, Senior Graphic Designer
Claire Benson, Associate Acquisitions Editor
Sean Adams, Project Editor
Luiz Ferreira, Senior Licensing Specialist

Copyright © 2017 by Cognella, Inc. All rights reserved. No part of this publication may be reprinted, reproduced, transmitted, or utilized in any form or by any electronic, mechanical, or other means, now known or hereafter invented, including photocopying, microfilming, and recording, or in any information retrieval system without the written permission of Cognella, Inc.

Trademark Notice: Product or corporate names may be trademarks or registered trademarks, and are used only for identification and explanation without intent to infringe.

Cover image copyright © iStockphoto LP/Jui-Chi Chan.

Printed in the United States of America

ISBN: 978-1-5165-0907-2 (pbk) / 978-1-5165-0908-9 (br)

CONTENTS

Introduction vii

HINDUISM 1

Introduction 1

Ecology 5
    Article 1 - *Hinduism and the Transforming Affect of Devotion* 6

Politics 19
    Article 2 - *Internet Hindus: Right-Wingers as*
    New India's Ideological War 20

Women 32
    Article 3 - *Marriage: Women in India* 33

BUDDHISM 39

Introduction 39

Ecology 43
    Article 1 - *Thailand: A Case Study* 44

Politics 57
    Article 2 - *Jhu Politics for Peace and a Righteous State?* 58

Women 68
    Article 3 - *Buddhist Nuns: Changes and Challenges* 69

JAINISM 80

Introduction 80

Ecology 83
Article 1 - *"They'll Know We are Process Thinkers by Our…" Finding the Ecological Ethic of Whitehead through the Lens of Jainism and Ecofeminist Care* 84

Politics 93
Article 2 - *The Morality of Sallekhana: The Jaina Practice of Fasting to Death* 94

Women 106
Article 3 - *Construction of Femaleness in Jain Devotional Literature* 107

JUDAISM 115

Introduction 115

Ecology 119
Article 1 - *Judaism and the Science of Ecology* 120

Politics 133
Article 2 - *The Transformation of Judaism in Israel* 134

Women 150
Article 3 - *A Worthier Place: Women, Reform Judaism, and the Presidents of the Hebrew Union College* 151

CHRISTIANITY 161

Introduction 161

Ecology 164
Article 1 - *Fundamentalist Dominion, Postmodern Ecology* 165

Politics 176
Article 2 - *Homosexuality* 177

Women 196
    Article 3 - *Sexual Awakening: Defining a Women's Pleasure* 197

ISLAM 208

Introduction 208

Ecology 212
    Article 1 - *Green Muslims* 213

Politics 222
    Article 2 - *Israel and Palestine* 223

Women 232
    Article 3 - *Restrictions of the Rights of Women* 233

Conclusion **251**

Credits **253**

INTRODUCTION

Trying to explain the concept of religion or how people study the subject of religion is like trying to pick apart a deeply connected organism. Many layers overlap each another to form a single complex system. At first glance, the system also appears to be a complete mess. To begin the process effectively, you need to start with one concept, so let's begin with this: Religions can act like languages. They give names, concepts, symbols, and a history to the sacred for communities around the world. These languages enable people to discuss the deepest aspects of themselves and the universe. The language is often difficult—and full of a symbolic communication that tries to bridge the regular and the ordinary with the sacred and the divine. In this way, religions delve into the seemingly unknowable, sometimes breaking boundaries and even reaching beyond the concepts of time and space. This religious quest usually provides concrete answers, at the same time that it leaves haunting questions or develops intricate philosophical ideas about who we are and why we exist.

Many scholars and theologians marvel at all the meaning that religion could give us, the cognitive structure and constructed realities that people tap into each day—especially as people try to process the coin-flip of life and death, death and life. It is common to feel unsettled about these notions or to back away from them, but for this school of thought, unfortunately, the rabbit hole goes even deeper. Other philosophers and scholars do not see religion as a means to understand beyond what we can see, feel, touch, and taste. For this second group, religion is a reflection of our own humanity. These thinkers process religion as a tool: a tool with which to see not just beyond but within. People have researched religion to understand how it functions within our psyche, within our practical societal structures, within our human behavior, and more. The list could seem endless. For example, is power yielded by religion, and if it is, is it used to legitimize violence? For some, religions simply hold up a mirror to show humans as communal creatures, as benevolent creatures, and as violent creatures. Therefore, we can study religion to understand our own humanity and not just what is beyond this physical world.

Once you have an overall view of religion, you can see how it can create deep meaning about the universe around us and how it can also shed light on the practical functions of everyday life. So how do you study that? The ways are countless. We can attempt to understand the difference between theological and academic approaches. For most of human history, religion was studied theologically, or at least it was studied through what would be the precursor to the theological approach. This earlier approach studies a religion as if every teaching in it were authoritative and valuable. It is considered a type of lens that gives the person studying the religion a faith-based perspective; that is, the student is taught to believe that each teaching about how the world works, and about the powers behind those inner workings, should be taken seriously. The interpretation of that authority can be drawn from a large spectrum. It can range from individual believers and spiritual leaders to whole institutions, or from a literal scriptural truth to a metaphorical one. It can be a combination of both—and anything in between. The importance lies in recognizing the spectrum.

The study of theology and its precursor is rich; in many cases, it is thousands of years old. But what happens for people who might want to study religion but are atheists or are agnostic? What about those who believe in one religion but want to study another? They all perhaps see value in studying the religion outside of its truth-claims. This approach will encompass how we will study religion in this anthology. It is known as the *academic approach*—or its precursor, the *comparative approach*—and it is only hundreds of years old. It began with technological advancements in travel and with world-changing inventions like the printing press. Then the study continued to develop, with large contributing factors like European colonialism and the 17th- and 18th-century movement called the Enlightenment. It is noteworthy to point out here that colonialism has a hard history of dominions plagued with killing, land grabs, and the tapping of resources. The world today still feels this impact as continents continue to struggle, many of them remaining underdeveloped and devastated by a legacy of corruption that touches all aspects of their political landscape.

However, somewhere in the middle of this colonial expansion we see accomplished a large feat: A world that seemed gigantic was made a tiny bit smaller. Encounters between continents that had been largely disconnected became common. The world became more open, and indigenous populations were exposed to Western belief systems. It also resulted in a European economic stability that created a class now able to engage in comparative religious study, with this same colonialism providing the examples for comparison. This impact was undeniable, especially as the comparative study of religion became systematic and entered the universities across Europe and Japan.

The comparative study of religion would also grow considerably within the period of the Enlightenment, which would inevitably leave its mark. Although we associate Western thought with the scientific method, the method's true genesis is found within Islam. A Muslim scientist known as Ibn al-Haytham (also called Alhazen) understood the need for this method hundreds of years before Roger Bacon wrote about it. This same method would be adopted and popularized during the Enlightenment. The Enlightenment sparked the huge paradigm shift that, for many people, replaced the basis of our knowledge: a shift from knowledge that was once mostly founded in religion to knowledge based on science. The study of religion would never be the same again. In fact, it might be a big reason for how and why we can assemble this type of anthology today. The collection of writings in this book underscores the larger questions of how and why religions play a role in our ecological concerns, political landscapes, and treatment of women.

Some scholars refer to this "new," post-Enlightenment approach as the *historical-critical method*. In it, religious study is historically contextualized and scrutinized. It is criticized in ways, and on a scale, that it might never have been before. This causes a dissection of religion; it can now be picked apart and broken down. The idea that religions do not happen in a vacuum has taken shape. Divinity and miraculous claims are all up for debate. The human factor, original manuscripts, and the culture in which a religion developed become details that must not be overlooked. However, the most important question about this new approach has been the following: What effect would this scientific scouring of religion have? From the beginning, the effect was reactionary, and it has been going on ever since.

This reaction resulted in extremist and fundamentalist movements across various religions. These movements placed an importance on returning to the very fundamentals of a given religion. Believers of all kinds, especially throughout Christianity, felt that their religions were being challenged—or even threatened with extinction. These movements still hold on to their religious knowledge tightly, sometimes sacrificing scientific knowledge in the process. I say "still hold" because these types of religious movements still flourish, evolve, and react to the new threats and attacks they feel. Ultimately, this is a critical concept to understand within each religion: the degree of social evolution. No single monolithic entity exists for any world religion. It can seem overwhelming at first, but there are branches, denominations, movements, sects, and even cults within all religions. And while these subgroups sometimes differ on the most important questions of salvation/liberation, they all ride under the banner of the same religion.

These broad understandings should start to paint the perfect picture of how complicated theology can be. By the same token, the academic study can be just as complicated. This stems from the fact that religion has no definitive boundaries. Religion cannot be put in a box and tied up with a ribbon and a bow. In fact, it would be impossible for theologians and academics alike to think of all the definitions of *religion* that have filled our collective human imagination. If religion was once a tool that our ancestors used to try to understand the world around them, today it has perhaps left us with more questions than answers. Unlike a mathematical problem, though, the academic study of religion has no formula with which to differentiate who or what is right or wrong. Today it is an interdisciplinary study that is easily analyzed from various vantage points. When we finally reach the bottom, what we have is a multilayered study that relies on the advancements and shortcomings of our previously established disciplines as well as the advancements and shortcomings of newly emerging ones.

Religion thus becomes one of the most difficult subjects to broach academically. It is also why this type of study has been constantly challenged and has been misunderstood time and time again. As a newer field of study, it still struggles to find its place in academia. However, a simple observation into the beginning of our 21st-century world informs us that we no longer can afford to undervalue the importance of such a study. More than ever before, the world is smaller, making this type of knowledge crucial if we hope to live on a shared, sustainable planet. That is the only way we can preserve a home for all of these different answers, questions, and philosophies that let us ponder what life and death are ultimately all about.

What better place to start our study than with ecology? Humans are indeed leaving a footprint, and the more we grow technologically, the larger that footprint becomes. It has now become so large that for many scientists the effects of climate change can no longer be stopped; they can only be slowed down. For this reason, climate change should be an issue of pressing and upmost importance. It can literally make or break our collective and continual survival on this planet, and religion should play a huge role in this conversation. Often, the way people perceive the natural world is informed by the belief systems they carry. Religious ideologies can therefore indirectly manifest into relationships with the natural world—or into the lack of such relationships.

This anthology in many ways is limited. Unfortunately, it lacks a conversation about indigenous communities. For thousands of years these communities held (and still hold) animistic beliefs. This means that in a sense the natural world is alive. Certainly, the natural world is animated: the rocks, the rivers, the animals, the trees all have a spiritual power to guide, heal, or harm. Indigenous communities have deeply rooted and intimate relationships with nature, and members of these communities feel related—as sons, daughters, and siblings—to many elements of the natural world. As a result, they have a deep reverence for nature. They protect nature and treat it as something precious. But how do other belief systems foster or discourage a close communing with the natural world?

In the following collection of articles, we'll specifically explore the two sets of closely related trinities: (1) the three Indian traditions of Hinduism, Buddhism, and Jainism and (2) the three Abrahamic faiths of Judaism, Christianity, and Islam. They all struggle against different environmental challenges. They also all manifest a wide spectrum of sacred teachings. These varied teachings in turn translate into many different relationships with the natural world. But within our world of globalized and unregulated growth, which parts of these relationships are being extolled and which are being shut out? This collection of articles will allow us to process these growing interactions between technology and the sacred. What are those relationships costing or fostering within environmental action or the lack thereof? It will also introduce us to *ecotheology*, or the idea that sacred teachings can have ecological consequences, whether it is an ecological ethic or an ecological action. On this subject, the topics explored include the Ganges River, deforestation, ecological ethics, Zionism, Fundamentalist Dominion theology, and Green Muslims.

But how religions can inform a relationship with nature is only part of the conversation. As implied above, these relationships can remain absent from the places where true application and practice can potentially impact billions of people. The boundaries and abuses that are established and allowed by policy and law across the multitude of governments around the world are key factors. Politics can mean real, day-to-day practice and interaction with the natural world. In fact, politics can mean daily practice and interaction with almost anything you could think of. A political conversation is therefore always essential, but when it is mixed with religion there is hardly ever any "black and white." Politicians can shut out religion as easily as they can use it. In fact, some governments are designed to be run by religious teachings and spiritual leaders. So if religion is a tool, it is today perhaps most easily seen being yielded within political landscapes.

The conversation about the relationship of religion to nature or religion to politics is far beyond the scope of this anthology, so introducing these ideas only provides a platform for future exploration. The articles included here allow us to process politics and religion in a specific number of countries, including India, Sri Lanka, Israel, and the United States. The collection of articles brings up a whole host of issues that intersect within the relationship between religion and politics. The topics include social media, monastic orders, morality, secularism, homosexuality, and suicide bombing. Presented alone, that list seems random and disparate. Yet it reflects the wide range of issues within the complex, grey world of religion and politics.

Lastly, this anthology explores religion and women. Although it is another topic that is far beyond the scope of this anthology, this relationship is just as complicated as the previous relationships. All the topics are complicated, because it is impossible to find one singular voice within any of these relationships. People can use religious scriptures to give a voice to the natural world, or they can use scripture to drown that voice out. People can also use religious scriptures to voice or attack political ideologies, just as easily as people use religious scriptures to empower or silence women. The relationship between religion and women is perhaps the most concrete, though. Many elements of the natural world do not have an audible voice, and many political systems have the amalgamated voices of too many. But what about women? Women can have amalgamated voices too, especially in movements, but you can always break it down to one single woman and her single story.

Ideally, this anthology would have collected the individual voices of women within their religions, respectively. However, these articles cover experiences and struggles for women across the six highlighted religions. The topics include marriage, nunhood, fasting, ordination, sexual pleasure, and restriction of rights. Women sometimes struggle similarly within many religions, but what can appear similar on the surface should not be confused with the specific lives these women live. Also, religion needs to be understood as only one facet of these struggles. Religions do not float up in the air aimlessly; they are anchored by people as religious leaders and followers, politicians and kings, etc. Religions can also heavily influence and be a part of the systems that are creating such intense struggles, most notably the environment of culture. Culture and religion are sometimes difficult to break apart, but in the articles in this book, exploring where religion and culture merge and where they separate is an essential analytical practice.

This anthology should provide only an introduction to, merely an idea about, what the relationship of religion to nature, to politics, or to women can look like. It should open your mind to understanding how religion is used and limited and how religion is merely one factor among many. Hopefully, it will give you an overview of the complete complex system I first mentioned.

I would like to conclude this introduction by stating my own limitations in the writing, processing, and understanding of religion. Whether it is on the study of religion or the teaching and history of each of the six individual religions, I am limited. I am limited in my worldview by the religion and culture I myself was raised into. I am also limited to my research and education. Having learned so much from religious communities, books, professors, colleagues, and students has allowed me to understand how much more there is to be learned. Although my intention is objectivity, I am sure my biases will show. Think, criticize, question, and analyze the following work as intensely as you can.

HINDUISM

INTRODUCTION

According to the sacred stories, the Ganges River came down from the heavens on Brahma's command. The impact of such a fall could have destroyed the entire world. But Shiva caught the falling river in his hair, breaking its fall and allowing its waters to run safely to Earth. Brahma and Shiva are two of the main gods who, along with Vishnu, make up the classical **trimurti**, or "trinity," within Hinduism. They are accompanied by a trinity of female counterpart goddesses: Saraswati, Parvati, and Lakshmi. But these are not all of the Hindu deities. In most images of Shiva, and although it can easily be missed, there is a streaming flow of water coming forth from his matted hair, one of the marks of an ascetic. It is the Ganges, not just a river but also a living goddess, that continues to play a vital role for the millions of Hindus who wish to commune with the sacred and be cleansed.

Understanding the gods and goddesses in Hinduism is no easy task. There are millions of gods, and yet many Hindus are able to *worship* only one god supremely, a concept known as *Henotheism*. In devotion to these gods, and despite the millions to choose from, some Hindus can also choose to *devote* themselves to only one. In the classical trinity we find Brahma, the god connected with creation; Shiva, the god connected with destruction; and Vishnu, the god connected with preservation. Yet Brahma has received the least attention from Hindu devotees, while Shiva and Vishnu—and the goddess Devi—receive a great deal. Devotion is practiced within the various temples as well as the frequent festivals in which the gods and goddesses are invited to temporarily inhabit a **murti**, or "image," usually a statue made of stone, metal, wood or plaster. It is then that devotees can engage in **darshan**, or "the act of seeing and being seen by the gods"; it is common to spot devotees doing their best to fervidly look into the eyes of the image.

The idea of "seeing beyond" is an important spiritual concept. Having an illuminated mind that sees the truth, the one reality, the gods, is not unique to Hinduism, but in the Hindu tradition it results in elaborate rituals. Devotees can engage in **puja**, or "worship," to show the gods honor. Honoring a god as a guest includes various rituals like bathing, dressing, and decorating the deity with garlands of flowers. Devotees also offer food that the gods partake of; what remains is considered **prasad**, or "blessed leftovers," which devotees can then eat. In classical Hindu creation stories, Brahma usually creates at Vishnu's command, and when Shiva destroys the world, he does so to beat it into a new existence. Some stories show variation depending on which god is regarded as the highest. These gods in turn also symbolize **samsara**, "the endless cycle of life, death, and rebirth." Within *samsara*, it can take millions of life-forms and lifetimes before a person becomes *moksha*, or "liberated." **Moksha** is the concept of liberation from *samsara*. A person's **atman**, or "soul/self," is what transmigrates, or literally wanders across to the next life-form and lifetime.

For many Hindus, this lifetime, the previous life, or the next one is the result of actions from previous lifetimes, all connected to each other as the person processes and fights against this cycle that is full of **maya**, or "illusion." **Karma**, which means "action," rests on the relationship of cause and effect. For many Hindus, devotion becomes *karma* that is eliminated by the gods. Because *karma* directly determines the form for a person's next life, sometimes maybe a life-form as uncomplex as a plant, for Hindus *karma* is a key ethical concept. The concept of *karma* was first introduced in some of the most ancient Hindu sacred texts, the *Vedas*. *Veda* literally means "knowledge." The *Upanishads* come at the end of what is known collectively as the *Vedic* literature; hence, they are called the *Vedanta*, the "end" (or "purpose") of the *Vedas*. It is in the *Upanishads* specifically that we are introduced to the concept of *karma*. The *Vedas* are today mostly known by the priests who study these sacred writings daily, in addition to consulting astrological charts.

The majority of Hindus, however, are connected to the more recent and classical sacred writings: the *Puranas*, or "Old Stories." Two of these are the long and great epic poems known as the *Mahabharata* and the *Ramayana*. Hindus connect to those epics and other sacred stories through plays, cartoons, soap operas, and festivals, to name a few. These sacred stories have some of the most famous avatars of Vishnu. **Avatars** are "incarnated forms of the gods." The sacred stories teach important lessons about the gods and about life. The *Bhagavad-Gita* (or *Gita*, for short) is a small part of the *Mahabharata* that is seminal and is found in most Hindu households. The *Gita* is essentially an intimate conversation between Lord Krishna (an avatar of Vishnu) and a young warrior named Arjuna. Krishna teaches Arjuna to do his duty as an act of devotion and without being interested in the outcome of his actions.

Let's examine the story a bit further: Arjuna was part of the warrior **varna**, or "caste." The warrior caste is the second of four castes, preceded by the priests and followed by the merchants/farmers and the commoners/laborers/craftsmen/artisans. The caste system, which traditionally excluded some people not only from the caste but from most aspects of society altogether, is controversial both inside and outside of India. Members of this group are often called "Untouchables." Even though Mahatma Gandhi's primary aim was not to address the plight of the Untouchables (he was even criticized by the leaders of some Untouchable communities), the poor condition of their lives did not receive global recognition until leaders like Gandhi forged movements to try to change these people's lives. Although there has been much reform that has allowed those outside of the caste system to be a part of society in many different ways, many families still hold religious and traditional values that rely on this system to determine their career paths, marriages, and even daily interactions.

This is because the caste system determines a Hindu's **dharma**, or sacred "duty." Fulfilling one's *dharma* creates good *karma* and helps a Hindu devotee reach a life-form and a lifetime closer to *moksha*. In the *Gita*, Krishna lays out various paths that Arjuna can take to reach *moksha* (although this this word is never mentioned in the *Gita*). The paths Krishna teaches include action, yoga, devotion, and knowledge. For many Hindus, to reach *moksha* means that their soul is liberated. Once that liberation occurs, believers may go to one of the heavens, or they may simply no longer participate in *samsara*. The *Gita* also harmonizes the seemingly disparate messages of the earlier and latter parts of the *Vedas*. The earlier parts emphasize many gods, as well as sacrifice; in the latter parts, the *Upanishads*, there is an emphasis on the one true reality of Brahman and meditation.

The teachings on meditation inside the *Upanishads* inspire the various yogic schools that are still in existence. In Hinduism, **yoga** is a form of meditation, a way to break down the body and allow it to serve as a vehicle for deep trancelike meditative states. Some *yogis* connect to a unified "superconsciousness of the universe or the divine," a meditative state also known as **samadhi**. Yoga also burns *karma*. In fact, in India today there are still **sadhus,** or "holy men/women," as well as *yogis*. *Sadhus* practice austere and punishing forms of yoga and asceticism. Most of them are devotees of Shiva, and they burn all of their material possessions and repress their sexual desires in pursuit of spiritual wealth. However, the *Gita* also teaches that devotees can remain active in society, showing devotion properly without needing to give up everything and become ascetics or a *yogis*. Devotees can be ascetics at home by being detached from the outcomes of performing their caste duties.

All of this diversity in Hinduism both reflects and influences the various cultures and languages of India itself. There is no uniform or universal system. This variety can be seen in all of the paths mentioned above, just like all of those paths that Krishna taught Arjuna in the *Gita*. Like all religions, Hinduism is not one monolithic entity, but there is perhaps a greater variety of practices and range of beliefs in Hinduism than there is in other religions. Across this huge spectrum of beliefs and practices, where no one is claimed as principal founder, many great teachers, gurus, and swamis have founded various paths. They all keep alive the ideal of spiritual learning and make spiritual wealth a priority over a material one. In addition, there are a number of different "ways of seeing" (six, traditionally speaking) or philosophical schools as well as yogic schools. Ultimately, the emphasis is on the "big picture." Hinduism encompasses a worship of the whole universe.

ECOLOGY

We have a systemic ecological crisis, and millions of Indians suffer from that impact daily. Some of the worst offenses are being felt by Indians who continually lose access to fresh water and to the food source provided by the marine life within most of their greatest rivers. The Ganges, although a sacred and living goddess, is the most polluted river of them all. The following article looks into the Yamuna River, the largest tributary of the Ganges. Millions of people die annually from ingesting the water from these polluted sources, and although globalization and development have been large contributing factors in this process, how can we understand the impact that the loss is also having on Hindu belief systems and practices?

ARTICLE 1

HINDUISM AND THE TRANSFORMING AFFECT OF DEVOTION

By John Grim and Mary Evelyn Tucker

The Yamuna River originates in a pristine area of the Himalayas at the Yamunotri glacier and at a hot spring in the region. Both of these sites have long been pilgrimage destinations. This revered river then drops down into the plains, where it meanders for 873 miles until it joins the Ganges at Allahabad and makes its way east to the Bay of Bengal. The religious relationships with the Yamuna at its glacial origins are different from those of the city of Vrindaban. In his insightful discussion of religion and ecology issues related to the Yamuna, David Haberman writes,

> All of the pictures of Yamuna Devi [goddess] … portray her seated on a lotus on the back of a turtle floating on a river, with high mountains in the background. … She is four-armed, holding a pot in her upper left hand, a lotus flower in her lower left hand, and a string of meditation beads in her lower right hand. … The symbols of the bountiful pot and creative lotus make it evident that Yamuna Devi is a powerful goddess who manifests life-giving forces and blessings.[3]

The river is ecologically robust at Yamunotri and is imaged as Mother Yamuna. Even guided fishing trips are promoted on this part of the Yamuna River, with images that present her beauty and her sacredness as connected to the vitality of fish and bird life.[4] As this religious imagery suggests, the river transforms those who come with reverent devotion to her.

The other source of the river is at a hot spring that gushes from a crevice in a massive rock face. This hot spring lies at the juncture of the Indian tectonic plate that has collided with the Eurasian plate for more than 70 million years. The force of the Indian plate slipping beneath the Eurasian plate has raised the Himalayas, the youngest and highest range of mountains in the world, generating the largest accumulation of ice and snow outside the polar ice caps. This tectonic collision also generated heat from contact with Earth's mantle. As water trapped in reservoirs deep in the crust of Earth is heated, it ascends through vertical channels to Earth's surface to generate hot springs, the signature of geothermal activity from shifting tectonic plates.

For the priests at the temple complex at Yamunotri there is another explanation for this thermal activity, a more poetic one that expresses the meaning of this source for the devout. The Yamuna is a living goddess who descended to Earth in answer to the prayers, austerities, and penance of seven great *rishis*, or sages. For many years the *rishis* resided at that pond and underwent austerities entreating her to come down to Earth to instill loving devotion among the people. Moved by their austerities and touched by their prayers, she descended to the summit of Mount Kalinda at 20,000 feet, but finding the place too cold for devotees to visit her, she prayed to her father, Surya, the Sun, to make the place more pleasant. With this request, the Sun gave his daughter a single fiery ray that struck the rock face at the base of the waterfall, giving birth to a hot spring.

Pilgrims bathe in this hot spring to commune with the goddess. The temple complex there is dedicated to devotion of the goddess Yamuna through her image as Mother. Her devotees consider the natural form of the river to be a more important manifestation of the goddess than her image. The goddess is embodied in the river; the divinity is the river herself. Therein lies a paradox. Some argue that it is only in the mountains where the river is free and clear that the goddess is truly alive, for now as the goddess descends to the plains her evident purity and holiness seem to be compromised.

The Yamuna as goddess is consistent all along her route as selfless and giving of her sacred waters of life. But the endless giving of the Yamuna has come up against the demands of a new worldview, that of industrialization, to build a modern India. Just as the religious devotion to the river changes along the route of the Yamuna, so the ecological and economic relationships with the river have changed with rapid development over the last 40 years. The question now is, Can there be reconciliation between the demands of tradition and modernity? Can religious devotion and economic development be mutually enhancing? Is the sacred in nature becoming replaced by the sacred in technology?

The Call of Modernity: Sacred Technology and Engineering

The Yamuna River, like all the sacred rivers of India, is a significant site for orienting and grounding religious practitioners in the flow of time.[5] As providers of nurturance and purification, these rivers are themselves places for transformation of mind and body. However, many of these functions, such as ritual and pilgrimage, have been subsumed by the allure of "sacred technology." Technology has become a means of transcending limitations and thus transforming the human condition.[6] In this sense, technology becomes a religious dream of liberating humans through modern progress.⁷ This is evident in the statement of the first prime minister of India, Jawaharlal Nehru: "Dams are the temples of a modern India regarding their capacity to provide hydroelectric power for massive production projects."⁸ Here we see that dams, not the rivers, are now considered sacred.

This engineering mindset has dominated India's political leaders as a way for the nation to enter into the modern status of an economically prosperous country. Much has been accomplished, but at great cost, and the Yamuna River bears witness to these costs. Dams have provided irrigation in the Doab region between the Yamuna and Ganges Rivers, but this management of the river has been almost totally oriented for immediate human needs. Rather than enhancing the flow of the river, it has radically altered and diminished it. Sustainable development has not been realized.

Guiding this engineering perspective is a bureaucratic mentality focused on centralized planning that largely ignores the human and natural communities along the river. This style of ecological management easily bypasses potentially creative local solutions regarding water usage in favor of a top-down model in which one size fits all situations. Thus, many residents remember interacting intimately with the Yamuna in religious rituals, recreation, and fishing into the 1970s. But centralized governance manipulated the river system so that these diverse local interactions were erased rather than incorporated into managing the river.

The situation in Delhi is instructive for understanding how centralized and provincial Indian governance has impaired regional water sources, knowingly and inadvertently. Lakes, natural catch basins, and wetlands along the river in the Delhi region have been largely drained for housing. Moreover, developers were allowed to use portions of the riverbed and the floodplain for buildings for the 2010 Commonwealth Games in Delhi. Before the Yamuna River reaches Delhi there is life in the river; after Delhi it becomes nearly impossible to sustain aquatic life. In Delhi eighteen drains channel urban waste directly into the river.⁹ Delhi officials recommend that even animals should not be cleaned in the river. Yet more than 20 million people live in Delhi and depend on the Yamuna for their water supply.

With population growth and rapid modernization, human demands for energy, drinking water, and waste management have increased in the last 40 years at rates that the Indian government was unable to anticipate. Moreover, the increased demand for food has led to massive irrigation projects and increased use of chemical fertilizers. The consequence is river diversion, diminished flow, and extensive pollution. Central government decision making has guided the types of agricultural practices chosen. For example, "green revolution" crops that held such promise have in fact required high chemical and water usage.

In 1993 the government of India launched the Yamuna Action Plan to clean the river, mainly through the construction of wastewater treatment plants located around the urban centers. Ironically, despite huge investments totaling US$308 million between 1993 and 2005, pollution levels have increased dramatically. With the huge expenditure of funds spent on the river and the lack of concrete results, the question arises why the approaches taken have not proved effective.

One response is that research on the Yamuna and implementation of recommended policies are fragmented on many levels. In the first place, studies of conditions of the river in different regions are generally unrelated to research in other regions of the river. For example, studies of river issues in the mountain regions related to deforestation are not connected to research concerning river issues on the agricultural plains. Investigations of urban issues such as wastewater management are not related to rural concerns for drinking water, irrigation, and fishing. Scientists examining the hydrology of the river tend to dismiss the initiatives of grassroots organizations for local watershed management. Second, writing and research are fragmented in terms of the disciplines that examine the river. Hydrologists use a different language from ecologists, and natural scientists speak a different language from that of social scientists in fields such as sociology, anthropology, and policy that also pertain to the river. Third, government policy has not been able to anticipate fully the tremendous population growth along the Yamuna River, especially in the urban centers such as Delhi, Mathura, and Agra. This population expansion has been matched by increased water usage by people and by industry. These developments have overwhelmed the wastewater treatment plants that the government has managed to create. Fourth, research and implementation seem to be divorced from the values concerning the river that are embedded in the religious and cultural traditions of Hinduism. Bringing these studies into relationship with one another is one of the motivations for religion and ecology as a field. Finding ways to bring local voices of civil society and religious communities into centralized planning for water conservation, tree planting, landscape restoration, and rain runoff management is one of the transformative long-term goals for religion and ecology. This type of civil advocacy on the part of religious organizations working on behalf of natural systems is not new in India, but new forms are emerging with policy implications.[10] Thus the conferences were organized in Delhi and Vrindaban.

Tension Between Development and Devotion

With the influence in India of modern values of efficiency and growth, the relationship with the Yamuna has become more exclusively an economic resource to be exploited than a sacred source that transforms devotees. Life along the river has had to accommodate massive engineering projects driven by new economic agendas. It is the combination of numerous dams, unregulated factories, and untreated urban waste that diminishes the flow and pollutes the river. Whereas religious pilgrimage and tourism represent ancient and enduring relationships between Hindus and the Yamuna, more contemporary industrial and agricultural uses have undermined these religious and ecological relations with the river.[11]

In 1991 a shift in official Indian development policy from the earlier Soviet-style socialism to Western-style capitalism did not alter the centralized planning or engineering mindset.[12] This shift was from a nationalist development ethos focused on large-scale state planning to more market-motivated expansion driven by competitive profit making and economic gain. The impact on the environment in both state planning and the market-driven development was deleterious and significantly increased with the latter. Flowing water, glaciers, mountains, minerals, soils, and biodiversity ceased to be manifestations of a unified reality evident in the ancient religious devotion to the river. To counter this economic exploitation, political rhetoric is invoked and gestures are made toward a religious appreciation of nature, but they rarely result in effective river cleanup. Some religious leaders, such as Srivatsa Goswami, have challenged the domination of economic profiteering, central planning, and massive engineering as the only acceptable approaches to treatment of the Yamuna River.

In subtle ways the older religious relationships allowed riverine biodiversity to flourish even as human populations increased.[13] This is evident in the Yamuna image at Yamunotri, where the presence of turtle, lotus, and mountains highlights awareness and respect for nature's beauty. It is manifest as well in the loving relationships of devotees to sacred forests along the Yamuna that are now mostly cut for development, except for small remnants left for tourism.[14] The centrality and value of biodiversity along the Yamuna reflect an enduring religious ecology within Hinduism. That is, it is understood that animal, plant, and human life occur in symbiotic relations with mountainous watersheds, mineral flows, and soil movements. This is an ecological vision that is present in the devotional practices found in Vrindaban, a vibrant pilgrimage city southeast of Delhi.

Devotion to Krishna and the River

Vrindaban is the setting for so many of the myths describing the life of the god Krishna. These stories of Krishna, and the religious ecology that has flowed out of them, suggest that in Hinduism the natural world is something beloved.[15] Krishna is considered by many Hindus to be the most significant *avatar*, or incarnation, of the supreme being, Vishnu.[16] In the stories of Krishna in Vrindaban, he plays (*lila*) with the Gopi maidens, the beautiful cow-herders who delight in him. Among the Gopi women, there is an intense devotion to Krishna and to Radha, his lover. Such devotion, known as *bhakti*, is a central religious practice of Hinduism, as seen in the *Bhagavad Gita* ("Song of the Lord").

Devotion to the god Krishna has resulted in an extensive religious ecology in Hinduism in which the natural world is interpreted as the manifestation of God's body.[17] Transformative images of the divine are believed to occur in the natural world. Natural objects such as trees, stones, or monkeys may manifest themselves as an icon of the divine (*murti*). For the devotee these natural manifestations are not the absolute fullness of the sacred but fitting expressions through which reverent devotion can be given to the sacred deity.

In the stories of Krishna, the Yamuna River is presented as a *murti*, or appearance of the sacred. Many of these mythic stories are collected in a twelfth-century religious text called the *Bhagavata Purana*. The religious practices, stories, and commentaries constitute expressions of a religious ecology that can be retrieved and reevaluated for transformative reconstruction of Hinduism in relation to the Yamuna.

The *Bhagavata Purana* gives the life story of Krishna at Vrindaban and describes his miraculous birth in the nearby city of Mathura:

> Then there was the supreme hour ... with all the stars and planets in a favorable position. Everywhere there was peace, the multitude of stars twinkled in the sky and the cities, towns, and pasturing grounds ... were at their best. With the rivers crystal clear, the lakes beautiful with lotuses and flocks of birds and swarms of bees sweetly singing their praise in the blooming forests, blew the breezes with a gentle touch fragrant and free from dust and burned the fires of the twice-born steadily undisturbed.
>
> The sages and the godly joyous showered the finest flowers and the clouds rumbled mildly like the ocean waves when in the deepest dark of the night Krishna, the World's Well-Wisher, appeared [was born] from the divine form of Lord Vishnu ... that wonderful child was resplendent with lotus-like eyes ... with yellow garments and a beautiful hue like that of rainclouds.[18]

The details are dazzling and highlight the cosmological character of the birth of Krishna, the World's Well-Wisher. The stars and deities, rivers and lakes, birds and insects all attend to and pay homage to his birth. The intimate relationships between Krishna and nature, especially the purity of the rivers, are evident. Throughout these stories, the love of Lord Krishna with and for the natural world is central. This is the *bhakti* that resonates between devotee and deity and is believed to be at the heart of all reality. In this religious tradition the heartfelt empathy for humans and nature is capable of transforming everything.

The numerous stories of devotion in the *Bhagavata Purana* tell how Krishna played as a child along the river and performed legendary deeds as an adult. There is a delightful human character to the stories of Krishna as a boy and culture hero. Krishna's divine and human character is layered in episodes that reflect dangers he faced. For example, in one story the demon Kaliya takes the form of a giant cobra who, motivated by revenge, completely poisoned the river.[19] In the mythic stories of the *Bhagavata Purana*, Krishna dives into the river and subdues the demon, dancing on his seven cobra heads until Kaliya cries out for mercy. This mythic incidence of toxicity in the Yamuna is seen by many as a remarkable prefiguring of contemporary pollution.[20]

These tales reflect the ancient mythology of Krishna, as lover and beloved, becoming one. In one of the famous hymns, an aspirant sings of this mystical vision yearning with devotion to partake in the river's embrace, the love bower itself. Hindu poet Chaturbhujadas extends this religious ecology into other symbiotic relationships in the natural world:

> Shri Yamuna favors her devotees and grants entrance into the love bower.
>> There Krishna, the Supreme Connoisseur of Love, makes love night and day.
>> To what extent can one describe that gathering of love?
>> Hearing Krishna's flute, the river stopped flowing and the women of Braj became enraptured.
>> No one can resist its sound.
>> Chaturbhujadas says: Yamuna is like a lotus,
>> My mind buzzes around her like a bumblebee.

As a mystical vision of union with the beloved god Krishna, these chants move the listener beyond divisions of the human from the divine. This religious ecology dispels sharp separations of nature from the human realm.

Indeed, this hymn reflects the devotion, *bhakti*, that strongly affirms the natural world as expressive of a larger unitive vision. This religious vision of the unity of all reality is present in the ancient literature of the *Upanishads* from the sixth century BCE. This oneness is named *Brahman*, and it is understood as residing within individuated reality as its *atman*, spiritual essence or self. In *bhakti* this experience of mystical union joins with emotional longing for the deity.[21] One undertakes song, dance, and gestures such as reverently laying a flower before a manifestation of the sacred or placing a drop of water on one's chosen deity. These devotional acts are considered comparable to the meditative acts of a yogi or the denial of worldly pleasures by a *sadhu*, or ascetic.

In this *bhakti*, devotees are not taught how to turn from the world toward a distant paradise or empty themselves of material attachments. Rather, the *Purana* teaches transformation by filling the practitioner with love of Krishna and thus becoming united with all reality.[22] In this vision, then, the devotee is welcomed into the love bower of the Yamuna River. In this loving interaction with the river, the pilgrim is affirmed by gazing on, bathing in, or sipping the water. This embodied knowledge provides insight into how devotion to natural systems enables transformation. That is, the practitioner forms deep experiential and symbolic bonding with such a natural feature as the Yamuna River. Traditionally, devotional practices have been largely personal, family, class, and caste related.

Throughout India these ancient practices are undertaken with sacred rivers, such as the Yamuna, Ganges, and Narmada. The water, or *jal*, of the river is understood as itself sacred and purifying of those who participate in its nurturance. For centuries, pilgrims have come to the rivers to express *bhakti* for personal transformation. Many Hindus come to die and be cremated along the banks as a final passage and purification for the next round of rebirth. These expressions of personal transformation are now being recognized to have implications for environmental civil advocacy, especially as the paradox of purity and pollution is becoming more evident. The overwhelming feeling of the Upanishadic unity of all reality and the goodness of nature experienced by the practitioner of *bhakti* now confronts the diminishment of the unity and the degradation of nature in the river.

Notes

1. For information on the conference "Yamuna River: A Confluence of Waters, a Crisis of Need," held January 3–5, 2011 at TERI University in Delhi and at Radha Raman Temple in Vrindaban, see http://fore.research.yale.edu/information/Yamuna_River_Conference.html; see also Richard Conniff, "The Yamuna River: India's Dying Goddess," in *Environment Yale* (Spring 2011):4–13; see http://environment.yale.edu/magazine/spring2011/the-yamuna-river-indias-dying-goddess/.

2. See Kelly Alley, "Separate Domains: Hinduism, Politics, and Environmental Pollution," in *Hinduism and Ecology: The Intersection of Earth, Sky, and Water*, ed. Christopher Chapple and Mary Evelyn Tucker (Cambridge, MA: Center for the Study of World Religions, 2000), 355–87; also see Kelly Alley, *On the Banks of the Ganga: When Wastewater Meets a Sacred River* (Ann Arbor: University of Michigan Press, 2002).

3. David Haberman, *River of Love in an Age of Pollution: The Yamuna River of Northern India* (Berkeley: University of California Press, 2006), 55–56. Haberman has studied the Yamuna for several decades and was key in the planning of the conference in Vrindaban in January 2011.

4. See http://www.india-angling.com/yamuna.html.

5. For policy approaches to the Narmada River, another sacred river in India, see Harry Blair, "Social Movements & Saving Rivers: What Can Be Learned from the Narmada?" published online at the Forum on Religion and Ecology website: http://fore.research.yale.edu/information/Yamuna/HBlair_NarmadaLessons.pdf.

6. Debate over the relationships of *technai*, arts and crafts, with *episteme*, knowledge, reaches back into both classical Sanskrit (*Dharmashastras*) and Greek (Plato and Aristotle) philosophical thought. A twentieth-century critic of technology was Martin Heidegger. He saw technology as a new transcendent axis that was intensified by the Protestant Reformation. Heidegger was deeply concerned with the linkage of science and technology as a way of perceiving the world. See Martin Heidegger, "The Question Concerning Technology," in *Philosophy of Technology: The Technological Condition: An Anthology*, ed. Robert Scharff and Val Dusek (Oxford, England: Blackwell, 2003), 252–64.

7. Thomas Berry wrote, "Once the scientific–technological period established itself . . . the intensity of its own dedication to its objectives took on the characteristics of a religious attitude and of a spiritual discipline parallel with the religious dedication and spiritual discipline of the classical religious cultures that preceded it. This included a new sense of orthodoxy, a new dogmatic integrity not to be challenged by reasonable persons. Yet . . . neither the creators of this new situation nor the spiritual personalities of the period have known how to read the change that has taken place." From "Christian Spirituality and the American Experience," in *The Dream of the Earth* (San Francisco, CA: Sierra Club Books, 1988), 118; Berry also observed that "scientific and social ideals have both become substitute mysticisms. Technology is the sacrament of our new birth. With their inner mystical dimensions and outer efficacy, science and technology provide an analysis of the human condition and a transforming

remedy." From "Traditional Religions in the Modern World," in *The Sacred Universe: Earth, Spirituality, and Religion in the Twenty-First Century*, ed. Mary Evelyn Tucker (New York: Columbia University Press, 2009), 11–12.

8. Thus, one of the first new, modern dams was planned on the Yamuna River at Dakpathar at the time of India's independence in 1947 and was completed in the 1960s. See C. V. J. Sharma, ed., *Modern Temples of India: Selected Speeches of Jawaharlal Nehru at Irrigation and Power Projects* (New Delhi: Central Board of Irrigation and Power, 1989).

9. See Central Pollution Control Board, *Water Quality Status of Yamuna River* (Delhi: Central Pollution Control Board, 2000); also Central Pollution Control Board, *Quality and Trend of River Yamuna (1977–1982)*, details at http://www.cpcb.nic.in/index.php.

10. For civil advocacy see Harry Blair, "Gauging Civil Society Advocacy: Charting Pluralist Pathways," in *Evaluating Democracy Support: Methods and Experiences*, ed. Peter Burnell (Stockholm: International Institute for Democracy and Electoral Assistance, and Swedish International Development Cooperation Agency, 2007), 171–92 and 228–39. See http://www.idea.int/publications/evaluating_democracy_support/upload/evaluating_democracy_support_cropped.pdf; see also Blair, "Social Movements & Saving Rivers."

    In this article, Blair observes that *civil advocacy* can be defined as organized activity not part of the state, the private sector, or the family, in which people act to promote mutual interests. In turn, *advocacy* can be defined as the process through which individuals or organizations endeavor to influence public policy making and implementation. Thus, *civil society advocacy* would mean efforts on the part of civil society organizations to influence state behavior.

11. When the Yamuna arrives on the northern plains of India from its origins in a Himalayan glacier, barrages and dams are the first set of controls imposed on the river. Barrages are simply lower dams used to direct the flow of the river. The Assan barrage at Dakpathar, finished in 1965, now redirects much of the water into utility canals for hydroelectric power and agricultural irrigation. Some would say that the river as a goddess ends here because for the remainder of its flow into the city of Delhi and beyond, the Yamuna is so diminished as to be no more than a managed canal.

    The Yamuna is further impeded at Tajewala by the massive Hathnikund barrage, where the river is broken into two canals that date from the Mughal dynasty of the fourteenth century CE. Often silted and cleared over the centuries, the Western canal now flows through the industrial cities of Yamunagar, Karnal, Panipat, and Sonepat. Industrial wastewater is pumped into the canals to replace the water extracted for irrigation of the large agricultural areas north of Delhi. The Eastern canal also supplies irrigation water for the "green revolution" agriculture area in the Doab between the Yamuna and Ganges Rivers. So much water is taken out of the Yamuna at Tajewala and directed into these two canals that in the dry season the main channel of the river is dry en route to Delhi.

12. The predominance of a managerial and engineering mindset in modern India gives little attention to sustainability of natural ecological processes. Rather, global economic development is mostly evident in websites and publications focused on human resource management among younger generations in India. See Pawan Budhwar and Jyotsna Bhatnagar, eds., *The Changing Faces of People Management in India* (New York: Routledge, 2009).

13. See Roy Rappaport, *Ecology, Meaning, and Religion* (Berkeley, CA: North Atlantic Books, 1979).

14. This devotion to the Yamuna River is also found in religions other than Hinduism, such as the respect shown by Muslim Mughal ruler Shahjahan, who built the magnificent Taj Mahal beside the river. This reverence appears in the poems of Muslim writer Mirza Ghalib and eclectic poet Kabir Das, and in the religious shrine along the river built by Sikh guru Gobind Singh. It was evident in fishing villages that are now long gone from the Yamuna River between Delhi and Vrindaban. There fish populations ceased flourishing in the 1980s, with deoxygenation and methane buildup resulting from urban waste flushed into the river.

15. This form of religiosity associated with the Mathura region of Uttar Pradesh (often called *Braj*) is practiced widely in northern India, including in Delhi, and in the states of Rajasthan and Haryana. The Vaishnava tradition of Hinduism also flourishes in southern India, and the relationships between the different regions are historically complex and intensely interactive. See William Sax, ed., *The Gods at Play: Lila in South Asia* (New York: Oxford University Press, 1995).

16. In Hinduism more than one deity can occupy the status of supreme being. That is, for the Vaishnava traditions, Vishnu is supreme, whereas for the Saivite tradition, Siva is supreme. These two positions need not be in conflict in Hindu theology, which allows multiple understandings of the unity of reality as expressed in the *Upanishads*.

17. See Milton Singer, ed., *Krishna: Myths, Rites and Attitudes* (Chicago: University of Chicago Press, 1966); John Hawley, *At Play with Krishna: Pilgrimage Dramas at Brindaban* (Princeton, NJ: Princeton University Press, 1981); *Songs of the Saints of India*, texts and notes by John Hawley, trans. John Hawley and Mark Juergensmeyer (New York: Oxford University Press, 1988); and Daniel Sheridan, *The Advaitic Theism of the Bhagavata Purana* (Delhi: Motilal Banarsidas, 1986).

18. *Bhagavata Purana*, canto 10, chapter 3 from *Srimad Bhagavatam*, online at http://www.srimadbhagavatam.org/.
19. This story is narrated in the *Bhagavata Purana*, canto 10, chapter 16 from *Srimad Bhagavatam*, online at http://www.srimadbhagavatam.org/.
20. See Haberman, *River of Love in an Age of Pollution*, 150ff.
21. Scholars have long noted the interweaving of *bhakti* dimensions in such texts as the *Isha Upanishad*. See *Upanishads*, trans. Patrick Olivelle (New York: Oxford University Press, 1996); and *The Thirteen Principal Upanishads*, trans. Robert Ernest Hume (Oxford, England: Oxford University Press, 1921).
22. Most importantly, social and economic divisions so crucial to the hierarchical society of India are subtly transformed in the heat of devotion. The mythic story describes the break from strict caste divisions by having all the wives of Prahlad, who are low-caste cow-herders, become the Gopi maidens who play (*lila*) with Krishna. These lower-caste women are the paradigmatic devotees.
23. See "A Tale of Two Rivers: Only One Happy Ending," *The Times of India*, July 8, 2012, http://timesofindia.indiatimes.com//articleshow/14738489.cms?intenttarget=no; "Delhi's Yamuna River: A Catastrophe in the Making," *The Hindu*, March 6, 2010, http://www.thehindu.com/sci-tech/energy-and-environment /article246228.ece; Rajar Banerji and Max Martin, "Yamuna: The River of Death," in *Homicide by Pesticides*, ed. Anil Agarwal (New Delhi: Centre for Science and Environment, 1997); and the Centre for Science and Environment website at http://cseindia .org/.See also http://indiatoday.intoday.in/story/yamuna-bachao-yatra-to-protest-at-jantar-mantar/1/257364.html.
24. It should be noted that in secular contexts this same kind of denial can be found about levels of pollution in rivers, wetlands, and oceans around the world, but rarely is the natural world set in the context of religion as clearly as it is in India. For example, the Po River, Italy's longest river, has experienced staggering industrial pollution and sewage from its urban centers, such as Milan. Only when the European Environmental Agency fined Milan was a treatment facility considered. The industrial and commercial benefits of using the Po as a dumping canal continued until environmental monitoring raised an alarm.

DISCUSSION QUESTIONS

1. How is the Yamuna River described? What is typical of her iconography?
2. How do Hindu beliefs about the Yamuna River help people develop a Hindu relationship with nature? What is that relationship like, and what type of ecological practices would it ideally promote?
3. Give concrete examples of the different factors leading to the pollution and ecological crisis of the Yamuna.
4. Can development and technology coexist with the sacred in the example of the Yamuna River? Why or why not?
5. How is Krishna connected to the river?
6. Do development and technology threaten Hindu belief and practice? If yes, what solutions could you provide?

POLITICS

There exists an ideological concept within India today coined *Hindutva*, and it literally means "Hinduness." This concept embraces Hindu nationalism and has been adopted by political parties as a right-wing platform. Although the majority of the people living in India are Hindu, many Muslims, Christians, Sikhs, Buddhists, and Jains are part of this overwhelmingly large population as well. There also exists a wide array of religious differences among them that ideally would produce very distinct political applications. The use of religion as a political tool is encountered in India as much as it is any other country. It is important to ponder the use of Hinduism within modern politics today. The following article navigates this precise concept in this era of social media.

ARTICLE 2

INTERNET HINDUS: RIGHT-WINGERS AS NEW INDIA'S IDEOLOGICAL WAR

By Sahana Udupa

We call ourselves the perfect nationalists

A HINDU RIGHT-WING TWEETER IN MUMBAI

With eighty million internet users on personal computers (24 percent of the population), thirty-nine million internet users on mobile phones (12 percent), and fifty-seven million on social media (17 percent), urban India constituted a growing community of online media users by the early decades of the new millennium. Needless to say, in a country like India, access to the internet implies that these users are economically privileged, and their use of social media implies considerable ease with the English language. However, with expanding smartphone markets and growing internet penetration, new online platforms for short messaging, microblogging, and social networking have made tremendous inroads across India.

Mumbai has the most social media users and the highest penetration rate for social media of any Indian city—a reflection of its large population and its importance as the country's economic powerhouse. According to the Indian Market Research Bureau, one of the largest market research companies in the country, the number of active online media users in Mumbai grew from 4.5 million in 2008 to 6.8 million in 2012—a significant 51 percent growth in just four years. By 2010, various social media platforms had entered the lives of online users in Mumbai, from Rise, Orkut, Facebook, and YouTube to LinkedIn and XING. After this first spell of social media expansion, the new microblogging site Twitter enticed online users with instant short messaging services along extended online networks, and WhatsApp modified this niche market to make it even more aggressive.

Twitter entered Mumbai and other major cities in India at a time when online users had grown familiar with and quite addicted to Facebook and LinkedIn, where they could reactivate, maintain, and expand their friendship groups, familial ties, and professional networks and satisfy, in some measure, their "insatiable nosiness about what other people are doing" (Miller 2011, 101). It was not surprising, then, that Twitter, despite its emphasis on imageless, short messages of no more than 140 characters, was initially used as an extension of personal networks, yet another online gateway for "real-world" socializing and snooping. Typically, people on Twitter would post updates on what they ate, what film they had watched, or where they went shopping. Naved, an avid online user from a Muslim neighborhood in central Mumbai, once told me, "A lot of people used it [Twitter] to say we are waking up, going to shower, just wore my shoes, which is really stupid!" Such updates on mundane, everyday routine conformed to the original design of Twitter as an online platform for sending messages to a small group of people with common interests and shared concerns.[1]

In just a year from Twitter's foray into the Indian market, tweeters in Mumbai grew enough in number to encourage some enterprising users to take the next step and propose off-line gatherings. In the middle-class neighborhoods of Chembur, Ghatkoper, and Bandra, with a large majority of educated Gujarati, Marathi, and Tamil residents, tweeters assembled for off-line tweet meets, endearingly calling them coffee tweet-ups and dinner tweet-ups. Some business-minded tweeters gathered to discuss budding technology projects. After such a gathering, they would typically go out to a movie or a restaurant and lounge chatting with their cyberpals till the wee hours. These closed-group tweeter meetings were driven by an assurance that they would be small, attended by those with common business interests or a shared cultural taste for movies and restaurants.

The terrorist attack on Mumbai in 2008 was a turning point. A medium couched comfortably for coffee table camaraderie became a potent site for organizing quick relief work. Shocked by the sudden crisis that had descended on the city, enterprising tweeters exchanged and coordinated a swell of online messages to reach out to people who were anxious to know the plight of their friends and family caught in the dramatic terrorist capture of four prominent locations in Mumbai. Tweeters with experience organizing tweet meets swung into action and spent hours on Twitter to connect the surging online queries with those for whom they were intended. They searched their lists of Twitter followers, alerted their friends on Facebook, or pulled up phone numbers from the network, creating a flow of information and connections that stood distinct from state channels and those opened up by mainstream mass media. This same voluntary relief work of informational service repeated during the floods of Mumbai in 2011, when tweeters managed a huge flow of queries and messages to arrange housing for people stranded at train stations or in their offices. After 11/26 and the Mumbai floods, a section of tweeters were convinced that they could achieve something with their own effort and knowledge of networks.

Twitter's transition from a platform for "silly updates" to one with the capacity to organize voluntary work—at least among English-educated middle-class groups during crises—was significant in the brief social trajectory of the microblogging site in Mumbai. By 2010, it had experienced yet another transition. Although routine updates and the impulse to organize tweeters for social causes or business interests did not disappear, Twitter came to be perceived more as a platform to share information, opinionate, and remain up to date on "hard" news. This transition coincided with Twitter's official branding in later years of its inception as a "real-time information network" where the user can "have access to the voices and information surrounding all that interests her/him."[2] The rapidly changing new media landscape and new smartphone applications were crucial for this branding, since Twitter had to distinguish itself from more fun-driven, frivolous, and at times outrageously intrusive new media services. Soon Twitter users commenting on politics and news events or tagging stories on "current affairs" were more common. Twitter became a high ground for opinion exchange, as opposed to what was increasingly seen as frivolous Facebook—a result of assigning "political moralities" to different media technologies (Miller 2011). Many online users in Mumbai had little doubt when they said that "Facebook is timepass; Twitter is for serious people." Despite the circulation of commentaries and "serious" content on Facebook, it was still largely seen as a platform for connecting intimate networks of friends rather than a public forum for serious matters. It was indeed the supposed public nature of Twitter, with its potential to open up channels of anonymous connections through pseudonymous and hidden handles (Twitter IDs), that confirmed its status as a "serious forum for serious people," as one social media user described it. In practice and in corporate branding, Twitter mimicked the repertoire of news and its claims to public opinion shaping, wrapped as such in tiny packets of 140 characters. What was more, like other interactive new media technologies, Twitter embodied an arena where recursive relations between virtual and off-line interactions were possible (Marshall 2001).

It was in this context of Twitter's market-mediated self-fashioning as a forum for opinion exchange that many celebrities entered the Twittersphere, to connect with the public and their fans without the mediation of mass media, and aspiring opinion shapers joined to post their comments and communicate directly with political leaders and cinema stars. Cybertalk trivia and market promotions continued on Twitter, but a section of politically savvy youth seized the opportunity to air their views without the filtering barriers of organized media, on a platform that promised to take them beyond tightly weaved online networks of friends and acquaintances. Such was the enthusiasm around this new short messaging service that the practice of *chalo ek tweet dalenge* (okay, let's just toss a tweet) became a sort of urban common sense. Yet this juvenile celebration of tossing a tweet and the *dikhawa* (display) of thumbing cell phones inherited Mumbai's deeply fractured political and cultural legacies: tweeting expanded in a megacity that was at once the quintessential symbol of postcolonial modernity and cosmopolitanism, the seat of India's dream factory, and a symbol of the very crisis of this secular developmental vision (Hansen 1999).

Excited by the possibility of directly connecting with mighty political leaders on a "publiclike" forum, a new generation of English-educated and technologically alert Hindutva sympathizers confidently plunged into the Twittersphere. In many cases, these sympathies were not preconstituted. As they tweeted and met with more tweeters online, they were drawn into the broad web of an emergent collective consciousness. Diverse as they were in their levels of ideological commitment, motivations to come online, and online style and poise, they nonetheless joined the growing group of online right-wing tweeters who engaged in Hindutva politics as a discursive practice. Some openly declared that they were right-wingers, self-christened themselves as "tweeple" (people who tweet), and used shared Twitter handles, notably "Internet Hindu," and similar handles and hashtags such as "The Proud Nationalist," "The Saffron Knight," and "Ex-Muslim."[3] This mission statement, for instance, flashes below the Twitter ID "Internet Hindu": "Reclaiming my motherland from Pseudoseculars. Bharata, the cradle of every other civilization and human existence."

By their own account, self-declared Internet Hindus are a growing community in the megacity of Mumbai, where I conducted ethnographic fieldwork among social media users in early 2012. "It is pure passion," an Internet Hindu who was barely in his twenties told me. "We don't get any money. It is passion—*kuch karne he* [we wanted to do something]." I was struck by the young man's enthusiasm and confidence; his energy evoked the alacrity of a youth leader. "Our vision is clear," he continued. "We are here to change the youth. Many youth Congress [party] people have joined us. [If] you follow us on Twitter, you will also be converted."

Internet Hindus in Mumbai imagine themselves as heroic warriors fighting an ideological battle on their own terms and upon their own will. To them, these platforms promise a completely autonomous arena where energies can cohere without any top-down mentoring or monitoring, where the youth can find a voice and a means of linking their voices, free of political might and manipulation. "First of all," an Internet Hindu emphasized, "we don't have an organization kind of a thing, with a secretary, president, and so on. It's only people and their passion. People are fed up with the system." A college student who had joined the discussion added, "There is a lot of frustration with people. We do this because we are frustrated with corruption." One finds here a deepening of the articulation of Hindutva with the liberalization discourse of the early 1990s (Rajagopal 2001),[4] nonetheless pitched against secular corruption after two decades of liberalization and two continuous regimes of the Indian National Congress party. This is in fact not entirely new or distinct from the strategies of organized Hindutva in India. As van der Veer (1994) and

Jaffrelot (1996) have astutely observed, Hindu nationalist politics has oscillated between ethnoreligious nationalism and socioeconomic issues of corruption and economic growth throughout its career in postcolonial India. On social media, these two discourses are brought together, erasing any possible contradiction between them.

The Allure of Anonymity and the Force of Hashtags

"We are the wheels to keep the communication running," an Internet Hindu declared in the quaint Mumbai neighborhood of Ghatkoper. "We are the wheels and the engine." The metaphor of wheels starkly contrasted with the image of leg workers of a larger organizer—wheels signified faster and more efficient movement and a sense of empowerment. To keep the communication running, most Internet Hindus assume pseudonyms and acronyms, often invoking grandiose images of the Hindu nation. For the most part, the romance of secret networks and underground rebels during the liberation movement has translated digital anonymity into imagined heroic politics of individual net warriors. This self-presentation unfolds in a network, an environment where anonymity is contingent on being inconsequential, since it can be disrupted with some effort to extract personal details, if this is found to be important and necessary. But despite the technological possibilities of decoding online traces and social networking sites turning into spaces where "time-space paths and patterns of interaction … become data points in algorithms" (Andrejevic 2013, 159), the network environment nonetheless provides an experiential sense of absolute anonymity. This is especially striking in the Mumbai context, which guarantees relative impunity to Hindu youth, who benefit from the historical bias of the Hindu-dominated police force against Muslims (Hansen 2001). Most Internet Hindus I met in Mumbai were quite assured of keeping their professional careers at a safe distance from their online identities. Varun, for instance, is a retail businessman who runs a small garment store close to his house in Kalyan, a large suburb. Born into a Marwari family, part of a caste group known for strong business networks across the country, he entertains customers from all religious groups, and some of his major suppliers are Muslims: "I am a businessman. As a businessman, I have to work with Hindus and Muhammadans. I want Twitter because I want to be me. I have a hidden handle because they [customers] should not think that 'kuch karta he, kuch aur sochta he' [he does something, thinks something else]."

In a savage twist of the virtual and the real, Varun finds in social media his true identity. The network environment not only guarantees experiential anonymity, which emboldens some of these

Internet Hindus to be their "real" selves, but also, in its continuous waves of interactions, ensures chance encounters among like-minded online surfers. Sundaran, a civil engineer who owns a construction company in Chennai, had traveled to Mumbai to attend a tweeters' gathering. He had tumbled into the world of microblogging before meeting any of the "patriotic nationalists" now assembled at the event: "Basically I came to know about Shiv [a fellow Internet Hindu] through Twitter. He started mentioning [a panic exodus of] northeast people and related topics. And then we connected. First we started, then our ideological mind-sets met . . . [and] that kind of [ideologically like-minded] people started following."

The Twitter field, akin to a magnetic field, brings together like minds with a force of attraction in what new media literature recognizes as "affinity spaces," where the affinity is to the ideological "endeavor and not other people" (Gee 2004, 84) and its creation can come from site designers and users alike.[5] The term *Internet Hindus*, then, suggests not any closely bound physical association but net users who cohere around common themes and issues in ideologically efficacious ways.

Many Internet Hindus admitted, in a tone of triumphant heroism, that their strategy in this effort is to first infuse clutter into the public discourse. As soon as an issue erupts into the public domain exposing the seemingly delicate relations between Hindus and Muslims in India, they raise a rapid wave of Twitter messages to drown out the "pseudosecular arguments" of the English-language media and later introduce a more coherent counternarrative by increasing momentum around a semantically rich, provocative hashtag. "PappuCII," a derogatory hashtag for the leader of the Indian National Congress, Rahul Gandhi, was one of many politically charged online efforts of propaganda designed and driven by Internet Hindus to debunk the party's claimed secular status. During several controversial events, including public rallies of Hindu nationalist leaders in Mumbai, Internet Hindus used shared IDs and together built hashtags to deluge social media platforms with provocative and abusive comments, confronting an equally abusive surge of comments espousing allegiance to Islamic radicalism. Such comments and hashtags kept alive a social media rumble on Hindu-Muslim animosity and hardened the view that Indian Muslims are active participants in international Islamic revivalism—a suspicion that was central to Hindutva activism after Muslim mobilization around the Shah Bano affair[6] and their association with the Iranian Revolution in the 1970s (Jaffrelot 1996, 339).

Significant in this revival of the key tenets of Hindu nationalism is the combination of the feeling of being vulnerable to "alien communities" (Jaffrelot 1996; van der Veer 1994) with newfound confidence around new media as tools for public arguments (or at least rabble-rousing) that can trump the mediation of organized news production. The new generation of e-savvy right-wingers have thus sharpened the discourse to associate Muslims with violence and threat, which is coeval with their strategy to deepen suspicion about organized English-language media, dubbed pseudosecular, hypocritical, and even blatantly antinational. Often, the producers of these media are derisively summed up by the slang term *libtards*—liberal retards—who are blamed for glossing minority appeasement and cowardly timidity as liberalism and secularism.

Internet Hindus are not always physically connected with the *shakhas*, the neighbor-hood organizational units, of the Rashtriya Swayamsevak Sangh (RSS), the parent right-wing Hindutva organization in India. Shakhas have been the locus of recruitment and ideological power for organized Hindutva, but Internet Hindus rarely attend their daily rituals of patriotic display and ideological grooming.[7] An Internet Hindu took pains to draw the distinction between Hindutva organizations and online Hindutva: "First, Twitter [as a particular form of online Hindutva] is not an organization. There is no hierarchy. It is a flat organization. No identity is required, actually. You don't need to be an RSS member. You say something good for the nation, society, and the individual, and you see people responding. People know me by my Twitter account and not my actual name. What matters is not identity but the content you post on Twitter or Facebook."

The absence of formal affiliation with Hindutva organizations suggests that Internet Hindus represent new forms of dispersed agency, which are inspired if not bound by the organizational authority of the RSS. For the RSS, the coalescing affinity spaces on social media are best left untouched—they can expand on their own and fight battles with their own arms. If the RSS has relied on its strategies to forge an alliance among "swayamsevaks, notables and men of religion" (Jaffrelot 1996, 351) on the one hand and continued, on the other hand, its social work and ideological grooming through local organizations, it has regarded the growing number of e-savvy right-wing Hindu youth as an army that both serves and disrupts its centralized organizational authority.

In an old yet affluently furnished building in Dadar, an expensive neighborhood in Mumbai, Pramod Bapat, the regional head of the RSS in Maharashtra, clarified his organization's position on Internet Hindus:

> Bapat: On [the] net, there are the doors. You can knock, you can enter, you can see, you can watch, you can use, you can get involved. [The] sangh does not plan separately or specially for social media. It [social media] is a social entity. It is a social bench. Everybody can sit there, everybody can talk, listen—that is why it is there! And thousands of *swayamsevaks* [volunteers] are working on that. They are using that media, they

have constructed very many communities and groups, they are sharing their thoughts, ideas, and working on causes.

Udupa: What causes?

Bapat: Like, say, love for the motherland. If this is the theme, what are the ways to express your love for the motherland? From media or such social platforms, they are spreading ideas, asking people to get involved, to know more about joining hands. But it is in their individual capacity [that] they are doing [this]. [The] sangh, as a whole and an organization, has not done such community [of "Internet Hindus"] or a page. Yes, we have an interactive website. The number of clicks on our website is increasing.

Partly since the RSS defines Internet Hindutva as a private endeavor carried out by individuals with no official approval or sponsorship, it has not issued a public call for restraint and moderation from Internet Hindus, signaling the deeply ambivalent status of this growing community. In his tiny office on a congested road of Bhendi Bazaar, widely seen as a Muslim ghetto, the Barelvi Sunni leader Saeed Noori lifted his hands and exclaimed, "Kya kare, Madam? Sab log fouji bang-eye" (What to do, Madam? Everyone has become a soldier). On a sprawling campus of the Hindu nationalist Bharatiya Janata Party's training center in Byandar, miles from Bhendi Bazaar and the crowded inner city, Vinay Sahasrabuddhe, a senior BJP leader active on Twitter, commented on his fellow tweeters, "You see, nobody comes with an empty head today."

The agency of Internet Hindus, and online political actors more generally, thus constitutes a dilemma for the organizational authorities, and its public visibility is always mired in uncertainties. One such moment of disruption and clutter arose when Internet HinduNDTV, a national English-language news channel, invited Internet Hindus to a talk show hosted by one of its prominent anchors. The channel had been trying hard to get these invisible online warriors to the studio and involve them in a debate on social media's highly effervescent forms of radical Hindu nationalism. It was not successful for a long time—many Internet Hindus told me that they firmly turned down the open invitation by the anchor on his shows. However, a few Internet Hindus were tempted. Their decision to appear on the show raised scathing criticism from a section of right-wingers online. A string of accusations and counteraccusations ensued. The confrontational repertoire hitherto reserved for online battles with "Muslim sympathizers" became the means of bickering within the community. Some right-wingers felt that representing Internet Hindus on mainstream television would rob them of their anonymity—affixing their agency within the limits of a corporeal body visible on television screens and exposed to asinine anchors. But this anxiety was couched in ideological terms. Online right-wingers asked their fellow Internet Hindus not to pay heed to the request of a "biased" channel that was, according to them, fully under "the control of the pseudosecular Congress party" and involved in "false propaganda." Even so, a few Internet Hindus braved the opposition from their camp and appeared on the talk show. Months later, one of them told me that his decision was based on "inside information" that the channel was planning to plant dummy Internet Hindus on the show. "Dummy and loose candidates would only [have been] bashed up by the opponents in the debate. This would have done more harm than some of us exposing our identity!" he said, justifying what others saw as a selfish act of hogging media publicity to the peril of collective anonymity.

These digitally mediated agencies, dispersed and diffuse as they are, and seemingly recalcitrant at times, rally behind related structures of authority. In online battles over issues from the territorial invocation of Akhand Bharat (Undivided India) and fears about a global conspiracy of Christian proselytization to the threat of multiplying Muslims and refusal to accept mosques as sacred places of worship, a set of "Twitter heroes" often anchor Internet Hindus. These internet icons tweet the most frequently and in terse, caustic, and provocative verse. Dr. Subrahmanian Swamy, an organizer of the Global Patriotic Tweeples Meet, for Hindu right-wing tweeters in Mumbai, is a *chanteur extraordinaire* of online Hindutva—an undisputed Twitter hero.

Long sidelined in formal national politics despite his key roles in the economic reforms of the early 1990s and the legal battles in the telecom spectrum scandal,[8] Swamy mastered the art of leading social media users through a tide of quick Twitter updates— posting and responding to comments, tagging links, and inventing an ever growing list of provocative acronyms to deride political and religious opponents. Soon he had amassed close to 270,000 followers on Twitter, with more than twenty-five thousand tweets in just two years. The newfound popularity of leaders like Swamy—at least among a section of net-savvy youth—what might be called net iconophilia, signals an emergent authority in the politics of religious difference in India. The key here is that they respond to individual tweets in a manner that evokes a sense of direct dialogue with ordinary people, exuding confidence that is unchecked and unrelenting. These leaders urge their audiences to pledge to continue their online activism and encourage them to invent new ways of adding force to their online heroism. Activism, after all, cannot remain in the virtual land forever. Twitter heroes regularly assemble crowds on the ground for tweeters' meets and get participants to reveal not just their names but also their full addresses—a feat that could turn social media marketing companies green with envy. The Global Patriotic Tweeples Meet was thus not a single episode but an itinerant gathering that carried the spirit of patriotic tweets to Mumbai, Delhi, Bangalore, Pune, and other major cities, with fervent calls on energetic youth to march ahead to "transition beyond Twitter." Such is the fascination with these new Twitter heroes that when I used the blank side of a copy of Swamy's speech at the meet, Viren, a self-proclaimed Internet Hindu who had traveled forty miles from a Mumbai suburb to attend the conference, gave me a scathing look. "So you are writing behind Swamiji's speech copy?" he chided; his towering figure made me quickly shove it down in my cramped bag and take out a disheveled notepad instead. The respectful term of address *Swamiji,* reserved for Swamy, signals ties of devotion similar to those enjoyed by a charismatic religious guru whose knowledge of history, politics, or theology is never open to challenge and never in doubt.

Many Internet Hindus in Mumbai told me that Narendra Modi from the BJP, who was contending for the prime ministerial position when I was conducting fieldwork and later emerged victorious, was "several notches above" Swamy in commanding social media allegiance, which is partly reflected in the number of his Twitter followers: more than two million. In a speech at the Global Patriotic Tweeples Meet in Mumbai, a professor of finance from the Institute of Management in Bangalore, a premier management school in India, described Modi as a modern-day Arjuna, the charismatic Pandava brother in the Mahabharata, and Swamy as Krishna, the shrewd uncle (and incarnation of Lord Vishnu) who ensured victory for the morally righteous Pandavas against the wicked Kauravas—a ready metaphor for the Congress party. The social media right-wingers gathered at the event responded with thundering applause as the culturally resonant Hindu mythological figures raised a wave of excitement in the audience. Ironically, the sense of net heroism as the work of individual ideological warriors obscures the structures of authority that Internet Hindus submit to and reinforce. Online icons such as Swamy illustrate this vividly. The energies were systematically channeled in later years, especially when BJP strategists set up "social media war rooms" for disseminating propaganda before the national elections in 2014.

While organized efforts to harness the energies of Internet Hindus picked up momentum, the new media architecture inspired practices that continued to mark online Hindutva's distinctness. For instance, Internet Hindus do not prize phases of political moderation in the discourse of the RSS or the politically pragmatic BJP or the general affability of *swayamsevaks,* which is doubtless partly a result of the privilege of anonymity. After a brief conversation with a leading Internet Hindu in Mumbai at a political meeting, I repeatedly attempted to contact him for a follow-up interview. In a black jumpsuit and shining black shoes to match, the young man in his twenties had looked more affluent than the modestly dressed rest of the audience. A badge for participants pinned on his suit gave his name, Atul, above his Twitter ID. I instantly recognized the ID, letting out an expression that prompted him to remark that he knew online Hindutva activists like him were being watched. Atul wavered between cordiality and downright disgust for an English-speaking academic who flew to India for flaneurlike visits. He would promise me all his support for my project and the very next minute would look at my notepad suspiciously and dismiss me as a pseudosecular academic with no credentials to speak about the lofty endeavor of Internet Hindus. In our brief, heated conversation at the meeting, I gathered that he owned and ran a social media marketing company in Mumbai. Afterward, I probed his whereabouts on the net and finally traced his cell number. Ready with questions and courteous words of reintroduction, I prepared to pitch an interview date. I was keen on this interview, especially since Atul's Twitter account regularly sent out the tweets of the Global Patriotic Tweeples Meet and Atul had proudly said he knew social media like the back of his hand. I doubted if he would grant an interview, but when I called, little did I expect to face a tirade on pseudoliberals and angry accusations that I knew too little about social media to merit a discussion with him. The exchange got worse when I evaded revealing my Twitter ID, which added to his suspicions about my intentions and, perhaps, his fears of disclosing his identity to a probing academic. He brusquely hung up the phone with one last piece of advice: "You post at least a thousand tweets and then come to me. You cannot understand us when you don't tweet." Shocked by these rude and scathing personal remarks, I turned to a friend-informant, Malini Sriram, a gender rights activist and among the most active tweeters in the city. Hurt as I was, I asked her whether it was true that one has to reach the mark of a thousand tweets before one can begin to understand Twitter. "It's the most silly, silly, silly thing," she retorted instantly. "It's like saying you cannot understand maternity problems without giving birth to a child." Herself subject to Internet Hindu intimidations on Twitter several times in the recent past, she told me, as a word of caution, "They are fanatics. They believe they are right and they are misunderstood. They use our own liberalism against us. I don't know if *liberalism* is the right word here—but after listening to them, we begin to feel 'Maybe I am wrong.'"

She consoled me that I was there to understand social behavior and not the technical details of a medium, adding, as a final word on the matter, that "Internet Hindus are engineers. They are very binary."

DISCUSSION QUESTIONS

1. Summarize the demographics of a typical Internet user within India.
2. The year 2010 saw an important transition in the use of Twitter for the residents of Mumbai. Describe the transition, and why it was so important for the political landscape.
3. How do the self-described "Internet Hindus" view themselves?
4. What is the Rashtriya Swayamsevak Sangh (RSS), and to what extent are the "Internet Hindus" connected to the RSS?
5. How was social media used to spread political propaganda, and by whom?
6. In detailing the differences between the campaigns of Swamiji and Narendra Modi in their efforts to become prime minister, what is the anecdote about the two candidates attempting to highlight? What are the implications of this anecdote?

WOMEN

It is difficult to think of a culture, a country, or a religion where women do not seem to have been handed the short end of the stick. Although Hinduism places a high regard on the feminine within concepts of the divine, Hindu women often have a set of very specific experiences that can be as beautiful as they are challenging. An Indian man was once quoted as saying, "In India we do not marry the woman we love, we love the woman we marry." Marriage in Hinduism is a great ordeal. There is a lot to consider religiously, especially as it impacts the family caste. How Hinduism molds marriage and the lives of the many women who become brides is a bigger topic than is encompassed in the following article, but this vignette gives us a small taste.

ARTICLE 3

MARRIAGE: WOMEN IN INDIA

By Doranne Jacobson

Hindu men ask about available mates for their children among their in-laws and relatives in other villages, and they discuss the virtues of each candidate with their womenfolk. Munni's father and uncles spoke with many relatives and caste fellows and heard of several prospects. One youth seemed acceptable on all counts, but then Rambai learned from a cousin that his mother had been widowed before she married the boy's father. Although widow remarriage is acceptable among members of Munni's caste, children of remarried widows are considered to have a very slightly tainted ancestry. Munni's parents looked further and finally decided that the best candidate was a seventeen-year-old youth named Amar Singh, from Khetpur, a village twenty miles away. He was the eldest son of a well-to-do farmer hitherto unrelated to their family. Munni's father's brother was able to visit Khetpur on the pretext of talking to someone there about buying a bullock, and he made inquiries and even saw the youth. Amar Singh had no obvious disabilities, had attended school through the fifth grade, and his family had a good reputation. Thus, after all in the family agreed, they asked the Nimkhera barber to visit Khetpur and gently hint at a proposal to Amar Singh's family. His relatives sent their barber to similarly glimpse Munni as she carried water from the well and to learn what he could about her and her family. Before too long, the fathers of the two youngsters met and agreed that their children would be married. A Brahman examined the horoscopes and saw no obstacle to the match. Each man gave the other five rupees as a gift for his child. Later, larger gifts were exchanged in a formal engagement ceremony.

In Bengal, a prospective bride may have to pass a rigorous inspection by her prospective father-in-law. At one such public examination, a village girl was tested in knowledge of reading, writing, sewing and knitting, manner of laughing, and appearance of her teeth, hair, and legs from ankle to knee.

> Rishikumar ... asked the girl to drop the skirt and walk a bit.
> The bride began to walk slowly.
> "Quick! more quick!" and silently the girl obeyed the order.
> "Now you see there is a brass jar underneath that pumpkin creeper in the yard.

Go and fetch that pot on your waist, and then come here and sit down on your seat."

The girl did as she was directed. As she was coming with the pot on her waist, Rishikumar watched her gait with a fixed gaze to find out whether the fingers and soles of the feet were having their full press on the earth. Because, if it is not so, the girl does not possess good signs and therefore would be rejected.[3]

The man read her palm, quizzed her on her knowledge of worship, demanded her horoscope, and asked that she prepare and serve tea. Even after he had found her acceptable and a dowry had been agreed upon, her bridegroom, eager to be modern, insisted upon seeing her—and could thus himself be seen by the girl and her people.

In cities, prospective marriage partners may exchange photographs, and the youth and his parents may be invited to tea, which the girl quietly serves to the guests. Each group assesses the other's candidate quickly under these awkward circumstances. Frequently, a girl is rejected for having too dark a complexion, since fair skin is a highly prized virtue in both village and town.

For city dwellers, matrimonial advertisements in newspapers often provide leads to eligible spouses. These advertisements typically stress beauty and education in a prospective bride and education and earning capacity in a groom. Regional and caste affiliations are usually mentioned.

> Required for our daughter suitable match. She is highly educated, fair, lovely, intelligent, conversant with social graces, home management, belongs to respected Punjabi family of established social standing. Boy should be tall, well educated, definitely above average, around thirty years of age or below, established in own business or managerial cadre. Contact Box 44946, The Times of India.

> Matrimonial correspondence invited from young, beautiful, educated, cultured, smart Gujarati girls for good looking, fair complexioned, graduate bachelor, well settled, Gujarati Vaishnav Vanik youth of 27 years, earning monthly Rs. 3000/-. Girl main consideration [i.e., large dowry not important]. Advertisement for wider choice only. Please apply Box 45380, The Times of India.

Discussions of dowry are important in marriage negotiations in conservative Hindu circles in many parts of North India. The parents of a highly educated boy may demand a large dowry, while a well-educated girl's parents may not have to offer as large a dowry as the parents of a relatively unschooled girl. In Central India, dowries are not important, although expensive gifts are presented to a groom.[4] In a few groups, the groom's family pays a bride-price to the girl's kinsmen. Almost all weddings involve expensive feasts, and the number of guests to be fed is sometimes negotiated.

As her wedding approached, Munni heard her relatives discussing the preparations. She pretended not to hear but was secretly excited and frightened. No one spoke directly to her of the wedding, but nothing was deliberately kept from her.

For Munni as for other villagers, her wedding was the most important event in her life. For days she was the center of attention, although her own role was merely to accept passively what happened to her. She was rubbed with purifying turmeric, dressed in fine clothes, and taken in procession to worship the Mother Goddess. Her relatives came from far and near, and the house was full of laughter and good food. Then excited messengers brought news of the arrival of her groom's all-male entourage from Khetpur, and fireworks heralding their advent lit the night sky. Munni was covered with a white sari, so that only her hands and feet protruded, and amid a wild din of drumming, singing, and the blaring of a brass band, she was taken out to throw a handful of dust at her groom.

The next day was a rush of events, the most exciting of which was the ceremony in which she was presented with an array of silver jewelry and silken clothing by her father-in-law and his kinsmen. Under her layers of drapery, the bride could neither see nor be seen but could hear the music and talking all around her. Many of the songs, sung by the Nimkhera women and female guests, hilariously insulted the groom and his relatives. Later, at night, in the darkest recesses of the house, her mother and *bhābhī* dressed Munni in her new finery. These valuable and glistening ornaments were hers, a wonderful treasure. Bright rings were put on her toes, a mark of her impending married state. Munni's little sister watched every ritual with wide eyes, realizing that one day she too would be a bride.

The wedding ceremony itself was conducted quietly at the astrological auspicious hour of 4 A.M. by a Brahman priest, before whom the couple sat. Amar Singh looked handsome in his turban and red wedding smock, but Munni was only a huddled white lump beside him. The priest chanted and offered sacrifices to the divine, and then, in a moving ritual, Rambai and Tej Singh symbolically gave Munni away to Amar Singh. As women sang softly, the garments of the couple were tied together, and the bridal pair were guided around a small sacrificial fire seven times. With these acts, Munni and Amar Singh were wed, and Munni officially became a *bahū*, a daughter-in-law and member of her husband's lineage.

As a *bahū*, she became a symbol of fertility, of promise for the continuation of her husband's family line. She also became an auspicious *suhāgin*, a woman with a living husband. The word *suhāgin* emphasizes the concept that neither man nor woman is complete as an individual but only in their union. Traditionally, no woman except a prostitute remains unmarried, and villagers believe that men who die single become ghosts who haunt the descendants of their more fortunate brothers. For some devout individuals, asceticism may be a stage of life, but except for a few holy men, all people are expected to marry.

The next day was another round of feasting, fun fests, ceremonies, and gift giving. As a send-off for the groom's party, Munni's kinswomen playfully dashed red dye into their faces. The relatives departed, the house was quiet. Only tattered colored paper decorations and her new jewelry served to remind Munni of her change in status. Life continued as before.

But Munni and Amar Singh, though strangers to each other, were now links between two kin groups, and their male relatives began to meet each other and become friendly. Munni's relatives were always properly deferential to Amar Singh's, as befitted the kinsmen of a bride in relationship to those of her groom. It would have been improper, too, for Munni's mother to meet Amar Singh or his mother or to speak to any of his male relatives. Having given a bride to Amar Singh's family, it now would be shameful for Munni's family ever to accept their gifts or hospitality.

Among Thakurs in North India, a girl is given in marriage to a boy who belongs to a group of higher rank than her own. Thus the bride's kinsmen are not merely deferential but are considered actually inferior to the groom's kin. This lower status of the bride's family adds to the relatively low status all North Indian brides have in their new homes. Among most Muslims, however, the kin of both bride and groom consider themselves equals, particularly since they often are close blood relatives (for example, the children of two siblings may marry).

The Muslim wedding consists of a series of rituals, gift exchanges, and feasts. The couple are legally united in a simple ceremony during which both bride and groom indicate their assent to the marriage by signing a formal wedding contract in the presence of witnesses. The groom and his family pledge to the bride a sum of money, known as *mehr*, to be paid to her upon her demand. (Most wives do not claim their *mehr* unless their husbands divorce them.) During the wedding the bride and groom sit in separate rooms and do not see each other. Latif Khan's mother was married by mail to her first cousin, living hundreds of miles away, and did not see her husband for over a year.

Munni expected to spend the rest of her married life as Amar Singh's wife, but she could remarry if he died. In her caste, as among most high-ranking groups, divorce is always a possibility, but it involves shame. For a Brahman girl, her first marriage would definitely have been her last until recently. Although most Brahman widows are expected to remain celibate for life, some Brahman groups now allow young widows to remarry without suffering ostracism. Among high-ranking Muslims, divorce is relatively rare, but it does occur, and remarriage is usually easy. In educated urban Hindu circles, divorce is almost unthinkable; but among tribals and low-status Hindus and Muslims, it is not uncommon, and scandalous elopements occasionally take place. In any case, a second marriage for a woman never involves the elaborate ceremonies of the first wedding but may simply entail setting up house with a new man.

During past centuries, very high castes prohibited widow remarriage, and a widow was sometimes expected to immolate herself on her husband's funeral pyre. This practice, known as *sati*, occurred in only a very small percentage of families and was legally abolished over a century ago. Rajputs remember with pride the *jauhar*, the rite in which the widows of warriors slain in battle died in a communal funeral pyre. Today, reports of *satis* appear in North Indian newspapers once or twice a year.

In the past, a widow was sometimes treated harshly, since the death of her husband was thought to be punishment for her misdeeds in a previous life. However, widows of lower-ranking castes have always been allowed to remarry. Under Indian law today, any woman may divorce her husband for certain causes, and any widow can remarry, but considerations of property and social acceptability rather than legality usually determine whether or not a woman seeks a divorce or remarriage. Among Hindu villagers, a widow who remarries customarily loses her rights in her husband's land. If she has young sons, a widow usually remains unmarried in order to protect her children's right to their patrimony.

In North and Central India, monogamy is generally practiced. Hindus may legally have only one wife, but Muslims are allowed four wives under both Indian and Muslim law. Village Hindus, whose marriages are seldom registered with legal authorities, occasionally take two or three wives, and the women of some Himalayan groups have several husbands.[5] Wealthy Muslim men occasionally avail themselves of their legal limit, but most cannot afford to do so. When Yusuf Miya, a Bhopal man, wanted to marry a second time, his wife spoke of suicide, and the matter was dropped. Some women, particularly those who have borne no children, do not openly object to having a co-wife. In Nimkhera, one untouchable sweeper man has four wives, all of whom contribute to his support.

Munni spent three more years in the bosom of her family, happily taking part in household and agricultural work and enjoying the frequent festivals observed in the village. Not long after her wedding, she went with her brothers and father to a fair in the district market center, where they watched the Ram Lila, a religious drama, and bought trinkets in the bazaar.

DISCUSSION QUESTIONS

1. What are the implications of remarriage for widows of certain castes?
2. Give details to describe the bride selection process undertaken by some Hindu families. What factors are highly considered when such a match is being made?
3. Describe the events and experiences that Munni lived through during her wedding ceremonies?
4. Traditionally speaking, do all women marry?
5. Is divorce accepted? If so, among which groups?
6. What is *sati*, and how has this tradition evolved?

BUDDHISM

INTRODUCTION

Buddha is a term that literally means "awakened one," and "waking up" is a core concept to all of the many forms of Buddhism. But what exactly are you waking up to? The concept could mean various things to various Buddhists at different stages of their lives and life-forms. Stated more simplistically, it could mean waking up to your mind and body—and understanding how undisciplined your mind and body truly are, whether that lack of discipline exhibits itself in the millions of mindless actions you take or the millions of thoughts that unknowingly cross your mind every day. You could be waking up to your present moment or even your true reality. Are you missing out on your life? What parts of it are you missing? These are some of the questions that might arise in the Western Buddhist practice. Waking up means awareness; it means perception. The underlying assumption is that possibly you are not awake—that you do not fully or properly perceive yourself, others, or the world around you. It also means that anyone can have the potential of waking up.

The aim of the various types of teachings on awareness and meditation, which are key practices among all forms of Buddhism, is to open the doors of perception: perception to the true nature of reality. Across Buddhist practice, the teachings about the nature of reality are deeply profound, and they deal with some of the world's most complicated philosophical discussions. The forms of meditation run a full spectrum of approaches and levels of difficulty. The basic concept of pushing against illusion should not sound foreign, however, as this is practiced, although differently, within Hinduism. Buddhism arises as a reaction to what we now think of as Hinduism, the older *Vedic* religion that was full of ritual and beliefs that left many feeling disconnected or hopeless. In many ways the worldview of Buddhism aligns with that of Hinduism. There exists a cycle of *samsara*, a "womb of rebirth," where there are countless lifetimes and life-forms to process. It was not until the Buddha's lifetime as **Siddhartha Gautama**, a prince born into the warrior caste of India, that he saw the full cycle of *samsara* and the process of rebirth.

Getting to the point of becoming fully awakened cost Siddhartha a lot. At the age of 29, he was pierced by the suffering in the world, and he compassionately gave up his luxurious life as a prince. This meant leaving his family, including his newborn son, and possessing nothing as he roamed the forests of India. He tried everything in an attempt to understand life and the nature of suffering; he outdid himself in yogic practices and then in asceticism, trying to find a solution. But it was all in vain. Siddhartha did not find what he was searching for until he looked inward and completely conquered **Mara**, the tempter god of desire. He did all of this deceivingly simply, as he meditated and stood his ground under the **Bodhi Tree**. For many Buddhists this tree, which was planted during the Buddha's time on Earth, still stands in the city of Bodh Gaya in northern India, and it is one of Buddhism's most sacred and central places. It was there that Siddhartha "awakened" and became the Buddha. It is important to note that for Buddhists the Buddha is not a god. Buddhists may recognize the gods of Hinduism, but they do not understand those gods or their roles in the same way that Hindus do.

Siddhartha's experience under the Bodhi Tree is referred to as reaching **nirvana**, or "enlightenment." Enlightenment awakened Siddhartha to see all of his past lifetimes and life-forms viscerally—the full cycle of *samsara*. He understood the nature of suffering inside this cycle, and he saw how we can release ourselves from it. Although the scriptures say that at first the Buddha was reluctant to teach, he taught until his death at around 80 years of age. In effect, he was the founder of Buddhism, and his teachings have impacted billions of people ever since. There are many forms of Buddhism today, and enlightenment means different things in the different forms of Buddhism. Each form has specific ways of understanding how and where enlightenment is attained. Some forms are found under the umbrella of *Mahayana* Buddhism, which means "the great vehicle" and dates to about the first century CE. This umbrella accommodates many branches of Buddhism across Buddhist Asia, including those in China and Japan.

The *Mahayana* traditions are unique to the cultures they evolved in. Each branch focuses on specific **sutras**, or "teachings." This is an important concept, because Buddhism has the most voluminous collection of sacred writings of almost any religion. This has occurred because the teachings come not only from the Buddha but also from other Buddha-like figures. Some of these figures are called **Bodhisattvas,** and many of the *Mahayana* branches have these figures. *Bodhisattvas* are savior-like figures that sacrifice themselves to stay inside of *samsara* after attaining enlightenment in order to help guide others. *Vajrayana* Buddhism, meaning the "diamond or thunderbolt vehicle," for some scholars dates to the sixth and seventh centuries CE. *Vajrayana* is sometimes seen as a part of *Mahayana*, and at other times it is viewed as its own form of Buddhism. This vehicle comes from Tibet, meaning it is unique to Tibetan culture, focused on specific scriptures but also having the idea of the *Bodhisattva*. Tibetan Buddhism is perhaps most famously known because of the **Dalai Lama**, who is a reincarnation of a *bodhisattva* known by the name Avalokitesvara.

The last form of Buddhism is referred to as *Theravada*, meaning "the way of the Elders." It originated under the **Hinayana** umbrella known as "the lesser vehicle," and today it is the only school remaining from this umbrella. It is primarily found in Thailand, Myanmar, Sri Lanka, and it is one of the oldest schools—probably originating sometime between 200 BCE and 100 BCE. It is one of the most conservative forms of Buddhism, following the teachings of the Buddha in the strictest fashion. Its teachings are canonized in what is known as the **Pali Canon**.

Despite the vast spectrum within Buddhist scriptures and practice, one of the core doctrines all Buddhists follow is collectively known as the Four Noble Truths. These truths summarize the heart and soul of the Buddhist perceptions of cause and effect and of impermanence. Although the last noble truth can have specific paths in different forms of Buddhism, the core message remains the same.

The first Noble Truth discusses one crucial element on the nature of reality. It asserts that the universe is permeated by **dukkah**, or "suffering." *Dukkah*, a word in the Pali language, is also sometimes translated as "disappointment" or "disillusionment." This suffering, disappointment, or disillusionment comes as the effect of an impermanent reality. This impermanence means a reality that is constantly in flux and imperfect. This constant change requires humans to let go of things as easily as they receive them; change and letting go are often some of the biggest challenges for humans. This reality makes nothing feel constant. The universe feels out of our control, and it is from this feeling that all the different forms of suffering, disappointment, or disillusionment arise.

The second Noble Truth establishes the cause of that suffering, and in the Pali language this cause is called **tanha**. *Tanha* means "thirst, craving, or desire." Not all desire is bad; what is bad is the desire that stems from our distorted vision of ourselves as separate entities.

So what is the vision of "self" that Buddhism teaches? The concept, called **anatta** in Pali, is actually that of "no-self" or "no-soul." Like many other religions, many forms of Buddhism break down the concept of the "self." This enables religions to attain compassionate and empathetic believers who are not selfish or "full of themselves." However, the Buddhist teaching of no-self does not mean that people are nothing. For many Buddhists, the illusion of constant self is really created by a combination of five *aggregates*, or different parts of what makes people feel like they are whole. Yet each part is constantly changing; the parts combine and recombine throughout a person's whole lifetime and life-form. In Buddhism, therefore, the "self" is fluid. The five constantly flowing aggregates include the material form, feelings, perceptions, will, and consciousness. So when the second noble truth discusses desire, thirst, and craving, the real problem is the desire that arises from thinking we have a self to desire for. There would be no problem, then, in desiring to become a Buddha, for example.

The third Noble Truth provides hope and establishes the idea that *tanha* can be stopped. The fourth Noble Truth lays out the path to how a follower can accomplish that. In *Theravada* Buddhism, there are eight ways, and this form of Buddhism is commonly referred to as the **eightfold path**. It includes fostering wisdom, virtue, and meditation through right view, right intention, right speech, right action, right livelihood, right effort, right mindfulness, and right concentration. Right view or understanding must be achieved first, before a person will be able to process any of the other paths correctly. The middle path, or the middle way, is another core teaching in Buddhism. In fact, it was one of the Buddha's first teachings. It involves finding balance by avoiding all extremes in life. (As Siddhartha lived with both extreme luxury and deprivation, this was a reflection of a lived experience for him.) Life experiences play an important role in Buddhism; they are to be considered even when we reflect on all of the teachings.

ECOLOGY

At the end of the 20th century, the rapid deforestation of the Amazon—which is commonly referred to as the lungs of the earth—was a wakeup call for many people. Today, the effect of the damage is still unfolding, especially as cities like Sao Paulo in Brazil enact water restrictions in response to devastating droughts. Perhaps this wakeup call comes too late, as the whole second half of the 20th century saw a rapid trend of deforestation everywhere in the world, even throughout Buddhist Asia. The following article analyzes the environmental crisis within Thailand. Between the 1940s and the 1980s, Thailand's forest coverage was reduced from 69% to 15%. Much of the human reaction to this devastation came from Buddhist monks.

ARTICLE 1
THAILAND: A CASE STUDY

By David L. Gosling

Buddhadasa Bhikkhu

Buddhadāsa Bhikkhu ('the monk who is a servant of the Buddha') was born in 1906 in Chaiya in southern Thailand. His mother was Thai, his father was a second-generation Chinese businessman, and his original name was Ngeuam Panich. He describes the primary influences on his life as his mother, the local *wat* and nature.

At the age of ten the young Buddhadāsa (Thai: Putatāt) became a *dek wat* (temple assistant) for three years, during which time he learned to read and write. He became familiar with temple life and collected medicinal herbs for the abbot. He ordained as a monk at the age of twenty and went to Bangkok to undertake Pali studies, but he was unhappy with city life and returned to Chaiya where he founded Suan Mokkhabalārāma ('the garden of the power of liberation') in 1932. The originality of his views soon began to attract attention, and many distinguished visitors came to see him. He gave lectures in Bangkok, teaching in a rational and incisive manner which appealed to the educated middle classes. He died in July 1993.

The cornerstone of Buddhadāsa's beliefs is that only emptiness or the void (*śūnyatā*) truly exists; everything else has a qualified reality—a view with strong similarities to the Mādhyamika philosophy from which the Mahāyāna stream of Buddhism developed. All existence is composed of transitory, impermanent events, but *śūnyatā* never changes; it is absolute being, absolute truth, *nirvāṇa* and the body of essence of the Buddha.

If ultimate reality is unchanging, then the cycle of rebirth (*saṁsāra*) cannot be a temporal process leading to *nirvāṇa*; it must be here and now, like *nirvāṇa* itself. Consistent with this interpretation, *anattā* (no-self) can now be regarded as a statement about the removal of self-centredness—the cause of attachment and consequent suffering—by wholesome actions. Wholesome actions and their consequences are determined by *paṭicca samuppāda*, or interdependent co-arising, which embraces all life and non-life in a web of interdependence.

Buddhadāsa's this-worldly 'here and now' interpretation of cardinal Buddhist doctrines is particularly appealing to busy professionals who have no time to meditate, and feel inferior to those who have. For them, *śūnyatā* represents a quality of activity in which the mind is tranquil, integrated and non-attached, as in meditation, though not through any meditative technique. In a personal interview with Buddhadāsa at Suan Mōkh in 1974, I asked him how similar this is to *niṣkāma* karma (work without attachment to its fruits) in the Bhagavadgītā.[7] He acknowledged the similarity, but pointed out that in Buddhism the 'empty' or 'void' mind becomes progressively 'nibbāned', or cooled, which is the etymological meaning of *nibbāna* (in a more temperate climate our hearts might be 'warmed'!).

Mahāyāna influence on Buddhadāsa is variously attributed to his Chinese father and the fact that the Mahāyāna tradition has always been strong in southern Thailand. The museum at Suan Mōkh contains many illustrations of *bodhisattvas*. There is an oil painting of the leaders of the world's major religions greeting one another, and a Zen cartoon of a group of people fastened together by a cord; the caption reads 'Human arrangement—by flowers'.

Buddhadāsa was too shrewd to become actively involved in politics, either to the right or to the left. He often spoke about '*dhammic* socialism', however, and maintained that there can be no ultimate separation between the spiritual and the social. *Dharma* (Pali: *dhamma*) is one with *dhammajāti*, or nature, which is the sum total of reality: 'The trees can speak, the rocks can speak, the pebbles and sand, the ants and insects, everything is able to speak.'[8]

Donald Swearer has studied the later years of Buddhadāsa's life, the period in which he became increasingly concerned about the destruction of the natural environment. According to Swearer, the elements of Buddhadāsa's ecological hermeneutic are to be found in his lecture at Suan Mōkh in 1990, in which the two central terms are 'care' (Pali: *anurakkhā*; Thai: *anurak*) and 'nature' (Pali: *dhammajāti*; Thai: *thamachāt*). The first of these terms is often loosely translated into English as 'conservation', and monks who oppose tree felling are described as 'forest conservation monks' (*phra kānanurak pā*). However, when Buddhadāsa uses this term he gives it a much more empathetic significance:

> One cares for the forest because one empathizes with the forest just as one cares for people, including oneself, because one has become empathetic. *Anurak*, the active expression of a state of empathy, is fundamentally linked to non-attachment or liberation from preoccupation with self, which is at the very core of Buddhadāsa's thought ... It is just such non-attachment or self-forgetting—the heart of the *dhamma*—that we learn from nature. We truly care for our total environment, including our fellow human beings, only when we have overcome selfishness and those qualities which empower

it: desire, greed, hatred … Caring in Buddhadāsa's *dhammic* sense, therefore, is the active expression of our empathetic identification with all life forms: sentient and non-sentient, human beings and nature.

Caring in this deeper sense of the meaning of *anurak* goes beyond the well-publicized strategies to protect and conserve the forest, such as ordaining trees, implemented by the conservation monks, as important as these strategies have become in Thailand. This is where the second term, *thamachāt*, enters the picture. The Thai term *thamachāt* is usually translated as 'nature'. Its Pali root, however, denotes everything that is linked to *dhamma* or that is *dhamma* originated (*jāti*). That is to say, *thamachāt* includes all things in their true, natural state, a condition that Buddhadāsa refers to as 'norm-al' or 'norm-ative' (*pakati*), that is, the way things are in the true, dhammic condition. To conserve (*anurak*) nature (*thamachāt*), therefore, translates as having at the core of one's very being the quality of empathetic caring for all things in the world in their natural conditions; that is to say, to care for them as they really are rather than as I might benefit from them or as I might like them to be. Indeed, *anurak thamachāt* implies that the 'I' is not over against nature but interactively co-dependent with it. In other words, the moral/spiritual quality of non-attachment or self-forgetfulness necessarily implies the ontological realization of interdependent co-arising.

From an ethical perspective this means that our care for nature derives from an ingrained selfless, empathetic response. It is not motivated by a need to satisfy our own pleasures as, say, in the maintenance of a beautiful garden or even by the admirable goal of conserving nature for our own physical and spiritual well-being or for the benefit of future generations.[9]

Buddhadāsa's view that *dharma* embraces the social and environmental spheres of human activities has inspired a number of progressive monks, though it is important to recognize that he did not encourage them to undertake community development activities themselves because this was the role of lay people. Among his chief followers are Phra Depvisuddhimedhi (also known as Phra Paññānantha), abbot of the Wat Cholapratan Rangsarit, and Phra Payom Kallayano, abbot of the Wat Suan Kaew (both in Nonthaburi). Phra Payom, who was a student of Buddhadāsa for seven years, spices his sermons with street-level slang and is very popular among young people. A typical off-the-cuff comment to me about monks' development programmes was as follows: 'If a monk is concerned about how to solve social problems, then he will not have time to think about removing any skirts.'

Phra Prayudh Payutto (also known by various honorific names such as Phra (or Chao Khun) Rajavaramuni and Phra Dhammapiṭaka) is a great admirer of Buddhadāsa, but differs from him in several important respects. His monastic career followed the traditional trajectory of Pali studies and the position of deputy secretary-general of Mahachulalongkorn Buddhist University, where he pioneered development activities for young scholar monks. Donald Swearer contrasts the environmental views of Phra Prayudh and Buddhadāsa as follows:

> Phra Prayudh grounds his argument for the value of nature for religious practice in stories of the Buddha and the early disciplines found in Pali texts. Buddhadāsa also links nature and religious practice to spiritual realization but does so by using Suan Mōkh as his primary illustration rather than citing specific passages in canon and commentary. Phra Prayudh, furthermore, makes a strong appeal to reason. Unlike some Thai Buddhist environmentalists who encourage such practices as ordaining trees or the promotion of a tree deity cult to preserve a stand of trees, Phra Prayudh believes that modern Buddhists need to go beyond appealing to Buddhist values, such as gratitude and loving-kindness, and citing scripturally grounded stories of the Buddha and the early *sangha*, and should utilize scientific evidence to address global problems, such as pollution and environmental preservation.[10]

Tambiah attributes the resilience and creativity of Thai monasticism to the absence of colonization combined with the intrinsic character of Buddhism:

> Virtually at all levels of society the integral relevance of their religion for conduct is not in doubt ... This attitude is partly the result of the greater sense of intactness and continuity experienced by the Thai as compared with other Asian societies actually colonized by Western imperial powers. But it also derives from the intrinsic character of Buddhism itself—how its tenets relate, on the one hand, to the confident claims of positivist science and, on the other, to the concerns of the politico-social order.[11]

Buddhadāsa and, to varying degrees, other monks, have been able to straddle both these worlds. Towards the end of his life Buddhadāsa became seriously ill. Everybody, including King Adulyadej Bhumibol (Rama IX), became deeply concerned, and Thailand's best doctors were despatched to attend to him. Three choices were open to them: they could treat him at Suan Mōkh, move him to a nearby hospital, or bring him to the Sirirath Hospital in Bangkok, one of the best in southeast Asia. The hospital's director, Dr Prawase Wasi, a long-standing admirer of the monk, was asked to convey a message to him from the King, requesting him 'not to leave his body so that he can help to maintain the *sasana* (religion)'.

'You can ask,' responded Buddhadāsa, 'but it all depends on causal conditions. If there are factors that enable the body to live, it will. If not, it won't. Don't try to carry the body away to escape death.'[12] He remained at Suan Mōkh, and recovered. He died two years later, aged 87.

Social and Environmental Activities

In 1964 monks living in Bangkok began to take part in two kinds of outreach to the provinces. The Phra Dhammatuta programme started that year under the auspices of the Department of Religious Affairs (part of the Ministry of Education and run mostly by former monks). Its primary aim was to promote national integration by strengthening people's attachment to Buddhism. It had a strong missionary emphasis and has sent a number of monks each year to work abroad.

The Phra Dhammajarik ('wandering *dhamma*') programme was jointly sponsored by the Ministry of the Interior via the Department of Public Welfare and the *sangha* in 1965; its main thrust was to spread Buddhism among border and tribal people. Somboon Suksamran describes it as 'a kind of moral rearmament mission to the northern areas where the tribesmen live and have been threatened by subversion'.[13] Both these programmes had strong political overtones from the outset and were not particularly successful.

Independent development schemes, known as *dhammapatana* ('development through *dhamma*'), have been organized by individual *wats* such as the Wat Phra Singh in Chiang Mai and by the two Buddhist universities in Bangkok. The first training programme for monks at Mahachulalongkorn Buddhist University began in 1966, and by 1972 was sending about sixty trainees a year to work in the provinces.[14] The programme also brought monks from the provinces to attend two-month courses in Bangkok; these included training in community and rural development, sanitation and first aid, and public health. There was also a lecture course on ecology and the environment. Mahamakut Buddhist University, which is mainly for *dham-mayuttika* monks, ran similar programmes, initially jointly with Mahachulalongkorn.[15] After a period of training, monks went to poor provincial areas where they took part in schemes to construct roads, bridges, school and temple buildings, wells and toilets. They also helped to install water pumps and electricity lines. William Klausner has described the involvement of monks in these activities in northeast Thailand in graphic detail.[16]

Monks sometimes operate their own development programmes based on particular *wats*. In Chiang Mai province Phra Khru Mongkol Silawongs, abbot of the Wat Bupparam in Chiang Mai city, runs vocational training schemes for electricians, builders and architects. Monks are taught how to help hill tribespeople to dig wells and build roads, and women are trained in weaving, sewing, toy making, and fruit and flower growing. In an interview Phra Khru Mongkol refuted the notion that his development work was political or that he wanted to make converts to Buddhism. He did it for its own sake and because the Buddha taught his followers to work for the welfare and happiness of others.[17] Chao Khun Rajavinayaporn is a senior *dhammayuttika* monk (Phra Khru Mongkol is *maha nikai*), who is deputy abbot of the Wat Chedi Luang in Chiang Mai. He supervises schemes to train women, and helps hill tribes raise money for development work, which includes the use of biogas and improved cooking facilities.[18]

The involvement of monks in social and environmental programmes raises important questions about appropriate and inappropriate behaviour for a monk. According to the Pāṭimokkha (which is part of the monastic Vinaya) a monk may not damage a plant or dig the earth (which might destroy small living creatures), but there is no reason why he cannot saw a log if somebody else has cut down a tree. The 227 rules of the Pāṭimokkha do not apply to a novice, who is subject only to ten precepts. The rules are particularly important to Thai monks because the Vinaya played a major part in King Mongkut's *sangha* reforms. Most Thai are therefore well aware of them, and public opinion is extremely sensitive to what is and what is not appropriate behaviour for a monk. From time to time the national press erupts with a story about some monk who is behaving inappropriately; this happened in July 1978 when Phra Kittiwuddho, a controversial politically rightist monk, was discovered to have had a Volvo car smuggled into the country for his use.[19]

Public opinion is less critical, however, if it is clear that there is a positive reason which can be endorsed from a Buddhist perspective why certain technically inappropriate activities are necessary. Thus, for example, Phra Chamrun Panchan, former abbot of the Wat Tham Krabok in Saraburi (he died in May 1999, to be succeeded by his brother), discovered a herbal medicine which, if administered in a therapeutic community based in his monastery, is highly successful in curing heroin and opium addicts.[20] Many of these addicts are teenagers, however, and it would hardly be reasonable to treat only the young men and not the women. The monks therefore feel obliged to do many things which, according to the Pāàimokkha, bring them into an inappropriate amount of contact with women. They must also, from time to time, operate a sauna, clear up lay people's vomit and pursue absconders. However, all this technically inappropriate behaviour may be condoned because the Buddha preached against the use of intoxicants. A person who cures drug addicts is therefore doing what the Buddha would have approved.

In 1976 Dr Prawase Wasi, director of the Sirirath Hospital and an eminent haematologist, arranged a three-week course on healthcare for monks at the Wat Thongnoppakun in Thonburi. This was followed by shorter five-day courses for groups of up to fifty monks at other Bangkok *wats*. These courses included the prevention and diagnosis of illness, childcare, and the treatment of illness using inexpensive traditional and modern medicines.[21] I have conducted studies of these courses and their effectiveness, in one case with particular reference to the availability and use of herbal medicines. Appendix A gives a list of the medicinal plants I photographed at a number of *wats* and subsequently identified from a catalogue in the library of Chulalongkorn University in Bangkok. The catalogue, by Ratdawan Boonratanakornit and Thanomchit Supawita, is called *Names of Herbs and their Uses*. It bears no date, and it could not be found anywhere else.[22] Appendix A lists the uses of each plant as specified in the catalogue.

I have also conducted studies of healthcare possibilities for monks and *mae chii* (lay nuns) in urban situations.[23] These *maw phra* (doctor-monk) schemes have been highly successful and have done a great deal to enhance public respect for the *sangha* and its lay supporters. Monks who practise basic healthcare are often dubbed 'bare-headed doctors'.

The social and environmental programmes described so far have aroused little controversy, but more recent attempts by monks to stem the tide of deforestation have led to major confrontations with the authorities. Trees have been ordained by encircling them with saffron cloth to prevent them from being felled, and sacred groves have similarly been created with sacred thread. Phra Thui from the Wat Dong Sii Chomphuu in Sakon Nakhon province has protected trees in this manner, and Phra Prajak Khuttajitto, a monk living in Dongyai forest in Buriram province in the northeast, was sent to prison for encroaching on a forest designated for the felling of trees.

The imprisonment of Phra Prajak occurred because of his support for poor farmers who were being resettled to make room for eucalyptus reforestation under a controversial resettlement programme set up in 1990 by General Suchinda Kraprayoon. Critics discovered that the real reason beneath the official veneer of environmental concern was that the army wanted to make money from private plantation companies. Phra Prajak was arrested in April 1991 for encroaching on forest reserve land and again in September, but in July the following year a new prime minister, Anand Panyarachun, abolished the resettlement programme.

Many of the social and environmental activist monks receive encouragement from Sulak Sivaraksa, who coordinates what has come to be known as 'engaged Buddhism'. Sulak is a Sino-Thai originally from Bangkok, who graduated in law in Britain. Following a distinguished literary and publishing career, during which he wrote extensively about the renewal of society through Buddhism, he attracted international attention in 1991 as a result of reactions to his comments about the monarchy. His remarks, made during a lecture at Thammasat University, could be construed as criticisms of the King, and he was arrested on charges of *lèse-majesté*.

Sulak believes that international capitalism and the consumer culture are primarily responsible for undermining Siamese society (he does not like to call it Thai!):

> The great department stores or shopping complexes have now replaced our *wats*, which used to be our schools, museums, art galleries, recreation centres and cultural centres as well as our hospitals and spiritual theatres. The rich have become immensely rich, while the poor remain poor or even become much poorer ... Not only our traditional culture, but our natural environment, too, is in crisis.[24]

Sulak's Buddhism owes much to Buddhadāsa. Thus he regards *nirvāṇa* not as a metaphysical reality but as an experience beyond the limits of the mundane: 'inner freedom, equilibrium, peace, void of angst and a sense of being entirely "at home" and unthreatened in the universe, which expresses itself both in a positive affective state and in compassion for all forms of life'.[25]

Sulak was eventually acquitted of the *lèse-majesté* charges, but was arrested in 1997 for taking part in direct action against the construction of the Yadana Gas pipeline in Kanchanaburi.[26] Environmental groups oppose this construction because it will destroy virgin forests in Thailand and encourage the military regime in Burma with its repressive policies for relocating villagers on the Burmese side of the border. Tree ordinations, attempts to lie down in front of trucks and appeals to the international community are all part of the campaign. The struggle continues ...

However, Buddhism can be a vehicle for environmental improvement in less dramatic ways. In education, for example, there are experimental school programmes which utilize Buddhist principles to promote community-oriented energy- and resource-efficiency schemes. One of these is based at the Wat Tongpuboran-Khanissorn municipal primary school in Ayutthaya. It began in 1997 as a collaborative venture between the Ministry of Education, the National Energy Policy Office and the Thailand Environment Institute, and is known as the DAWN project.

According to Khru Surin, one of the school's teachers,

> Our attempts to jump onto the economic expressway have landed us in disaster. Look at the vast rice fields … In a rush to get rich quick, farmers have poured in so many toxic chemicals to boost yields that they've killed the fish in the ponds as well as put their own health at risk.[27]

The project manager is Dr Uthai Dulyakasem, former dean of the Faculty of Education at Silpakorn University. He believes that all education must be geared to the Buddhist notion of right (or perfected) understanding (the first step on the Noble Eightfold Path):

> Education must lead towards Right Understanding, which in Buddhist teaching is *sammā-ditthi*. The future generations must realize that saving energy is not a personal matter but a vital concern of the entire society … The DAWN programme hopes to enable people to see themselves as responsible for the entire process, from production to consumption. This means that they must be able to see and connect things from a holistic perspective.[28]

A reform in the entire learning process is essential for the creation of a better future; it must change people's patterns of consumption while at the same time nurturing their sense of social responsibility. The school curriculum draws upon the resources of the local community and integrates environmental matters into every subject, using them to explore the links between academic disciplines usually taught in isolation from one another. Thus, for example, a discussion of how much energy has been utilized in the production, transportation and consumption of a commodity could form part of a course in geography, physics or mathematics, and a class on Buddhism could be used to analyse the manner in which commercial advertisements stimulate the desires and illusions of unbridled consumption.

Notes

1. A Thai monk should be greeted by folding the hands under the chin and lowering one's eyes. The hand gesture is known as a *wai*.

2. Michael J. G. Parnwell and Raymond L. Bryant (eds) *Environmental Change in South-East Asia*, London, Routledge, 1996, p. 6. More generally, see also Philip Hirsch and Carol Warren (eds) *The Politics of Environment in Southeast Asia*, London, Routledge, 1998.
3. Jonathan Rigg (ed.) *Counting the Costs: Economic Growth and Environmental Change in Thailand*, Singapore, Institute of Southeast Asian Studies, 1995, p. 13.
4. Stanley J. Tambiah, *World Conqueror and World Renouncer*, Cambridge, Cambridge University Press, 1976, p. 185.
5. Ibid., p. 102.
6. Ibid., p. 401.
7. David L. Gosling, 'Thai Buddhism in transition', *Religion*, 1977, vol. 7, no. 1, pp. 18–34.
8. Santikaro Bhikkhu, 'Buddhadāsa Bhikkhu', in Christopher S. Queen and Sallie B. King (eds) *Engaged Buddhism*, Albany, State University of New York Press, 1996, p. 160.
9. Donald K. Swearer, 'The hermeneutics of Buddhist ecology', in Mary E. Tucker and Duncan R. Williams (eds) *Buddhism and Ecology*, Harvard, Harvard Center for the Study of World Religions, 1997, pp. 26–7.
10. Ibid., p. 34.
11. Tambiah, op. cit. (note 4), p. 429.
12. Ampa Santimetaneedol, 'Facing death with dignity', *Bangkok Post*, 6 November 1991.
13. Somboon Suksamran, *Political Buddhism in Southeast Asia*, London, St Martin's Press, 1976, p. 104. See also, same author, *Buddhism and Politics in Thailand*, Singapore, Institute of Southeast Asian Studies, 1982.
14. David L. Gosling, 'New directions in Thai Buddhism', *Modern Asian Studies*, 1980, vol. 14, no. 3, p. 415.
15. David L. Gosling, 'Thai monks in rural development', *Southeast Asian Journal of Social Science*, 1981, vol. 9, nos. 1–2, pp. 78–85.
16. William J. Klausner, *Reflections on Thai Culture*, Bangkok, Siam Society, 1993.
17. David L. Gosling, 'Redefining the saṅgha's role in Northern Thailand: an investigation of monastic careers at five Chiang Mai wats', *Journal of the Siam Society*, 1983, vol. 71, parts 1 and 2, p. 94.
18. David L. Gosling, 'Biogas for rural development: transferring the technology', *Biomass*, 1982, vol. 2, no. 4, pp. 309–16. See also, same author, *Vitritakan Palangngan nai Prathet Thai le Asiatawanogchiengtai* (Energy Crisis in Thailand and Southeast Asia), Bangkok, Komol Keemthong Foundation, 1981.

19. Gosling, op. cit. (note 14), p. 431.
20. David L. Gosling, 'Visions of salvation: a Thai Buddhist experience of ecumenism', *Modern Asian Studies*, 1992, vol. 26, no. 1, p. 37.
21. David L. Gosling, 'Thailand's bare-headed doctors', *Modern Asian Studies*, 1985, vol. 19, no. 4, p. 793. See also, same author, *Maw Phra* (Doctor-Monk), Bangkok, Komol Keemthong Foundation, 1986.
22. David L. Gosling, 'Thailand's bare-headed doctors: Thai monks in rural health care', *Journal of the Siam Society*, 1986, vol. 74, pp. 83–106, and in shortened form in *Journal of the National Research Council of Thailand*, 1987, vol. 19, no. 1, part II, pp. 1–10.
23. David L. Gosling, 'Thai monks and lay nuns (*mae chii*) in urban health care', *Anthropology and Medicine*, 1998, vol. 5, no. 1, pp. 5–23.
24. Donald K. Swearer, 'Sulak Sivaraksa's vision for renewing society', in Queen and King, op. cit. (note 8), p. 210.
25. Ibid., p. 222.
26. 'Sulak steps up pipeline campaign', *Bangkok Post*, 2 August 1998, p. 1.
27. Vasana Chinvarakorn, 'Dawn of a new age', *Bangkok Post*, 'Outlook', 6 December 1999, p. 1.
28. Ibid.
29. David L. Gosling, 'Urban Thai Buddhist attitudes to development', *Journal of the Siam Society*, 1996, vol. 84, part 2, pp. 103–20.
30. Charles F. Keyes, 'Political crisis and militant Buddhism in contemporary Thailand', in Bardwell L. Smith (ed.) *Religion and Legitimation of Power in Thailand, Laos and Burma*, Chambersberg, Pennsylvania, Anima, 1978, p. 149.
31. 'Scholar demands action over Dhammachayo letter', *Bangkok Post*, 1 May 1999.
32. Peter A. Jackson, *Buddhism, Legitimation and Conflict*, Singapore, Institute of Southeast Asian Studies, 1989, p. 212.
33. Phichai Tovivich, 'Monosodium glutamate: poisoning food for profit', in David L. Gosling and Feliciano V. Cariño (eds) *Technology from the Underside*, Geneva, World Council of Churches and Quezon City, National Council of Churches of the Philippines, 1986, pp. 87–92.
34. Khin Thitsa, *Providence and Prostitution*, Change International Reports, London, Parnell House, 1980, p. 4.
35. Nanthana Chaiyasut (ed.) *Report on a Survey of the Status of Women in Two Provinces*, Bangkok, publisher unknown, 1977, p. 26.

36. Chatsumarn Kabilsingh, *Thai Women in Buddhism*, Berkeley, Parallax Press, 1991, p. 19.
37. Ibid.
38. A. Thomas Kirsch, 'Economy, polity and change in Thailand', in G. W. Skinner and Thomas Kirsch (eds) *Change and Persistence in Thai Society,* Ithaca, Cornell University Press, 1975, pp. 172–96.
39. Thitsa, op. cit. (note 34), p. 23.
40. Charles F. Keyes, 'Mother or mistress but never a monk: Buddhist notions of female gender in rural Thailand', *American Ethnologist,* 1984, vol. 11, part 2, p. 224.
41. Kabilsingh, op. cit. (note 36), p. 25.
42. Nancy J. Barnes, 'Buddhist women and nuns' order in Asia', in Queen and King, op. cit. (note 8), p. 261.
43. Susan Murcott, *The First Buddhist Women,* Berkeley, Parallax Press, 1991, p. 10.
44. Kabilsingh, op. cit. (note 36), p. 36.
45. Samer Boonma, 'Bhikkhunî in Buddhism', unpublished MA thesis, Bangkok, Chulalongkorn University, 1978, pp. 114–25.
46. *The Rules of the Foundation of the Nun Institute of Thailand,* Bangkok, Suntsiri Press, 1979, p. 35.
47. David L. Gosling, 'The changing roles of Thailand's lay nuns (*mae chii*)', *Southeast Asian Journal of Social Science,* 1998, vol. 26, no. 1, pp. 121–43; see also, same author, op. cit. (note 23), p. 133.
48. David L. Gosling, 'Buddhism for peace', *Southeast Asian Journal of Social Science,* 1984, vol. 12, no. 1, pp. 59–70.

DISCUSSION QUESTIONS

1. What does the following statement mean, and what are the ecological implications of it? "The trees can speak, the rocks can speak, the pebbles and the sand, the ants and insects, everything is able to speak."
2. Describe the type of care provided by "forest conservation monks."
3. The idea that a monk cannot damage a plant or dig in the earth is an example of what?
4. Give examples of how monks try to preserve the trees.
5. According to the article, what is the impact of International Capitalism?
6. Explain the vision and goals of the DAWN project.

POLITICS

For most of Sri Lankan history, monks were absent from active political participation. Walpola Rahula's book *The Heritage of the Bhikkhu*, published in 1974, helped to change that. It was published just as a new Sri Lankan Buddhist political movement came into existence. Rahula described the roles of monks in the public sphere and as political advocates. This paved the way for a more recent event: In 2004, nine monks became official politicians in the Sri Lankan parliament. This move echoed throughout Southeast Asia, and people are still trying to process this overlap. The following article discusses the National Sindhala Heritage Party, or JHU, a political party made up solely of Buddhist monks.

ARTICLE 2

JHU POLITICS FOR PEACE AND A RIGHTEOUS STATE?

By Mahinda Deegalle

Monks in the JHU Election Platform

Establishing a Buddhist state (Sin. *Bauddha rājya*) in Sri Lanka is the main objective of the monks of the JHU. In their political agenda, the highest priority is given to the determination for a Buddhist state. Devout Sinhala Buddhists are also keen to see this happen since they are fed up with moral decadence and chaos that has emerged in contemporary Sri Lanka.

On the whole, five reasons can be identified as motivating factors that led the Buddhist monks of the JHU to contest in the general election held in April 2004:

1. the perception of Venerable Sōma's untimely death as a systematic conspiracy to weaken Buddhist reformation and renewal,
2. increasing accusations of intensified 'unethical' Christian conversions of poor Buddhists and Hindus,
3. continuing fears of the LTTE's Eelam in the context of recent peace negotiations,
4. the unstable political situation in which the two main political parties—UNP and SLFP—are in power-struggle in the midst of resolving the current ethnic problem and
5. the political ambitions of some JHU monks.

Traditionally, the majority of Theravāda Buddhist monks have stayed away from politics. The monks of the JHU entering into Sri Lankan parliamentary politics is problematic both from cultural and religious perspectives. Due to the controversial nature of the issue and debates over monks' actions, the JHU monks themselves have tried to explain the current political and social circumstances that led them to take such an unconventional decision. Their entry into active politics, they consider as the last resort, 'a decision taken with much reluctance'.[49] Before handing over the nominations for April 2004 elections, Venerable Athurāliyē Rathana, media spokesman of the JHU remarked: 'the Sangha has entered the arena of politics to ensure the protection of Buddhist heritage and values which had been undermined for centuries'.[50]

Why did Buddhist monks decide to contest in the parliamentary elections? Their answer lies in the following justifications:

1. The first justification is concerned with possible political disadvantages that the Sinhala-Buddhist majority may face as a result of the current peace negotiations with the LTTE initiated by Norway facilitators. According to the JHU, popular consensus is that there is no more division within Sri Lankan society as pro-UNP and anti-UNP. The two prominent Sinhala dominated parties—the United People's Freedom Alliance and the United People's Front—stand for the same principles. To secure power within Sri Lankan politics, both parties are ready to negotiate with the LTTE on the ISGA (Interim Self-Governing Authority) proposals forwarded by the LTTE through Norway. From the point of view of the JHU, these negotiations may disadvantage the Sinhala-Buddhist majority.
2. The second justification is related to the current tense environment created by unethical conversions initiated by non-denominational, evangelical, Protestant Christian groups. A Buddhist group named 'Jayagrahajaya' (Success—Sri Lanka), founded in Kandy in 1991 and approved as a charity by the Sri Lankan government in 1995, has written extensively on 'unethical' conversions carried on among poor Buddhists and Hindus by various nondenominational Christian groups which the Success identifies as 'Christian fundamentalists'. To inform the public the threat that exists for Buddhism, the Success has also published a booklet: *Āgam Māruva* (Changing Religions) written by Venerable Mädagama Dhammananda (Dhammananda Thera 2001), Project Director of the Success. According to Success, there are over 150 NGOs registered in Sri Lanka under Company Registration Act who carry out conversions.[51]

Various Buddhist groups including the monks who formed the JHU have demanded from the Sri Lankan government to pass a bill in the parliament to ban unethical conversions carried out among poor Buddhists and Hindus. The current controversy with regard to unethical conversions is not purely a Buddhist concern. Before the current bills, former Hindu Affairs Minister T. Maheswaran had challenged the former Prime Minister Ranil Wickremasinghe that he would resign if 'the government did not bring in an act to prevent Hindus being converted to Christianity before the 31st of December' 2003.[52] Venerable Ōmalpē Sōbhita, JHU MP, and

Rajawattee Wappa sat on a fast outside the Ministry of Buddhasāsana on 30 December 2003 demanding the government to take an action on unethical conversions.[53] Newspapers captured 'smiling minister W.J.M. Lokubandara walked up to them through the group of well-wishers and onlookers and sought their permission to sit beside them'. Though Mr W.J.M. Lokubandara, Minister of Justice and Buddhasāsana, promised to take an action, he could not do anything since President Chandrika Bandaranaike Kumaranatunga dissolved the parliament in early 2004.

In this uncertain and inactive political context, the JHU believed that both major Sinhala-dominated political 'parties are not willing to ban unethical religious conversions'.[54] This loss of hope and frustration had led the JHU monks to decide to enter into the legislature. As a result of the JHU's demands, by June 2004, there were two bills on 'unethical conversions' in the Sri Lankan Parliament for approval. Venerable Ō. Sōbhita published his bill on 28 May 2004 and the Sri Lankan Government also drafted a bill. These legislative measures show how religious concerns have become important in private and public lives of Sri Lankans. Two important factors—conspiracy theories surrounding Venerable Soma's death and the potential threats of unethical conversion to Buddhists and Hindus—have motivated the JHU monks to enter into politics.

It is also possible to identify four key phases that mark significant mileposts in the gradual development of present political activism of Buddhist monks of the JHU by drawing support from a wide range of ideologies and a cross section of Sri Lankan population:

1. the founding of Jathika Sakgha Sabhāva (National Sakgha Council) in 1997 by drawing support from the monks of the three monastic fraternities,
2. the birth of SU (Sinhala Heritage) Party on 20 April 2000,
3. the birth of Jathika Sakgha Sammēlanaya (National Sakgha Assembly) and 4 subsequent formation of JHU in February 2004 as an all-monk political party to contest April 2004 election.

All these political movements, in one way or another, embraced an idealized notion of the *dharmarājya* (righteous state) concept thought be the underlying public policy of the ancient (Buddhist) polities of Sri Lanka. It was perceived that in the most authentic form, the *dharmarājya* concept was present in the government policies of Emperor Afoka in the third century BCE. The Buddhist monk politicians of the JHU capitalize this idealized image of *dharmarājya* concept for their own political advantage in contemporary Sri Lanka.

The *Dharmarājya* Concept of the JHU

To attract a captivating audience, the JHU has introduced more fashionable religious terms for its political rhetoric. One of them is the *pratipatti pūjāva*, which literally means 'an offering of principles'. The Sinhala term *pūjāva* is strictly speaking liturgical in its connotations and exclusively used in religious contexts rather than in the political platform. However, the JHU has employed it self-consciously in the highly charged expression *pratipattipūjāva* in order to introduce its political manifesto in religious terms connoting their ambition of establishing a *dharmarājya* in Sri Lanka.

The election manifesto of the JHU is rather unique because of its interesting religious content and the way it was introduced to the Sri Lankan public by invoking religious sentiments. Unlike other political parties, the JHU offered its political manifesto (*pratipattipūjāva*) to the Tooth Relic of the Gotama Buddha in Kandy.[55] On 2 March 2004, the JHU monks and lay supporters marched to the Tooth Relic Temple, Kandy from Kelaniya Temple[56] in the midst of thousands of Buddhist monks and lay people who shared the noble mission of restoring Buddhasāsana (message of the Buddha) and promoting Buddhism in Sri Lanka.

The selection of 2 March as the date of launching the election manifesto and Kandy as the place for the launch are quite significant in historic terms. March 2 symbolizes an important historic event: the day that Sri Lanka lost her independence to the British under Kandyan Convention signed on 2 March 1815. Another event that happened on that day in Kandy still in the ears of the Sinhala nationalists: when the British raised the union jack before signing the memorandum, the monk Kuḍāpola protested against it and he was shot dead there. The JHU's unveiling of its programme at a gathering in Kandy deliberately invokes religious and national sentiments.

The JHU launched its political manifesto in the hope of restoring the weakening status of Buddhism in Sri Lanka. The monks of the JHU have a clear agenda and ambition of purifying the political process from corruption and abuses. The JHU manifesto includes twelve points as principles for constructing a righteous state (Sin. *dharmarājyayak udesā vū pratipatti pūjāva*). Each item in the manifesto includes more than one principle and the Sinhala version[57] is more comprehensive than the English rendering.

1. The first principle stresses that Sri Lanka should be ruled according to Buddhist principles as it was in the past and the protection of the Buddhasāsana should be the foremost duty of any government.[58] The state is, however, identified in the manifesto as a 'Sinhala state'.[59] The state also should safeguard the rights of other religions to practice their own religious traditions. Showing the urgency of addressing religious

concerns of the majority and achieving political ambitions of the JHU, the very first principle of the manifesto mentions the issue of unethical conversions. It asserts, 'all unethical conversions are illegal'. This is an indication that the JHU will take a legislative action on 'unethical conversions' once its members are elected to the parliament.

2. The second article stresses that Sri Lanka is a Buddhist unitary state that cannot be divided.[60] National safety is an essential condition. At times when there are threats to national security, without political interference, the Police and the Three Armed Forces should be given powers to act according to the constitution to safeguard the national interests of the country.

3. Emphasizing the JHU's stand as National Sinhala Heritage Party, the manifesto states that national heritage of a country belongs to the ethnic group who made the country into a habitable civilization. The hereditary rights of the Sinhalese should be granted while protecting the rights of other communities who inhabit the island.

4. The rulers of Sri Lanka should adopt the *dharmarāja* concept of Emperor Aśoka, which was influenced by Buddhist Philosophy, and should work for the welfare of all ethnic groups. Their exemplary attitude should reflect Dharmāśoka's idea of 'all citizens are my children' (*save munisā mama pajā*).

5. The Government should control and monitor all the activities and monetary transactions of the non-government organizations (NGOs) that are in operation in Sri Lanka. This is an indication of a religious concern that the JHU has raised with accusations to evangelical Christians that the majority of NGOs that are registered in Sri Lanka under the corporation law undertake evangelical activities of converting poor Buddhists and Hindus to Christianity in the guise of providing technical education.

6. Following the *grāma rājya* concept that Sri Lanka inherited, a decentralized administration should be adopted. This is the Buddhist option that the JHU plans to adopt instead of devolution proposals that the successive Sri Lankan governments plan to implement to resolve the ethnic conflict that has arisen with terrorist activities of the LTTE. The JHU sees negatively the devolution of power as a solution to continuing ethnic problem in Sri Lanka. They maintain that the notion of devolution of power is an imported concept imposed upon them with vested interests to break Sri Lanka.[61] Their negative attitude to devolution of power is based on two factors: their fear that it will lead to the creation of a separate state for Tamils and it will lead to the creation

of fanatical religious beliefs and conflicts within Sri Lanka. Instead of the devolution of power, the JHU prefers a 'decentralization' within a unitary Buddhist state. They believe that effective 'decentralization' to village level communes will solve many of the issues related to defense, administration, education, health, trade, agriculture, water and transport. Their conception of 'decentralization', they identify as '*grāma rājya saṅkalpaya*'.

7. The development should centre on the natural habitat, animals and humanity. The development should be based on the principle 'by developing the individual human being, country should be developed'.[62] The creation of a just, national economy based upon Buddhist economic philosophy and empowering local farmers and entrepreneurs.

8. An education system that fits into the Sri Lankan cultural context and that meets the needs of the modern world should be introduced. A society in which the lay–monastic, male–female, employer–employee, child–parent, teacher–student, ruler–ruled who are mutually bound by duty should be introduced. A righteous society in which the five precepts are observed should be built on the basis of Buddhism.

9. In the past, Sri Lanka was the land of *dhamma*, which spread Buddhism around the world. Therefore, international relationships should be established with sister Buddhist countries. Friendships should be built with other countries. Whilst maintaining close relationships with the neighbouring countries, we should consider that Sri Lanka is an independent state.

10. A Buddhist council should be held to reinforce Sinhala *bhikkhu* lineage and the recommendations of 1957 and 2002 Buddhist Commission Reports[63] should be appropriately adopted.

11. Female moral rights, which are destroyed by commercialization, should be safeguarded. Nobility and dignity of motherhood should be restored.

12. Independent, free and ethical principles should be adopted for mass media.

These twelve points demonstrate guiding principles of the JHU as a Buddhist political party in Sri Lanka. In engaging in politics and in presenting this twelve-point manifesto, the key visible political motive of the monks of the JHU is their desire in creating a 'Buddhist voice' within the Sri Lankan Parliament so that Buddhist and Sinhala interests can be secured and guaranteed within the legislature. Increasingly, they perceive that power-hungry-Sinhala-lay-politicians have betrayed the Sinhala and Buddhist rights of the majority population of the country.

Election Victory, Chaos and the JHU in the Parliament

The JHU monks have faced several controversies within the JHU's short existence so far. The act of nominating over 200 monks to contest a parliamentary election was controversial in itself. The JHU's act of using monks to contest the election has been criticized both in abroad and Sri Lanka. In addition, the existing major political parties attempt to weaken them at every possible opportunity.

The election success of the JHU, however, was a shock for many who perceived their significance very lightly since none of the candidates were highly versatile politicians. In the election held on 2 April 2004, the United People's Freedom Alliance (UPFA)—a combination of SLFP and JVP—won 105 seats out of 225. The UNP, the Sri Lankan government from 2001–2004, was defeated in the election and secured only 82 seats. Tamil National Alliance (TNA) backed by the LTTE won 22 seats. The JHU and the Sri Lanka Muslim congress (SLMC) had 9 and 5 seats respectively.

As the newest political party, the JHU had a significant success in the election; though its candidates were novices to parliamentary politics, they were able to convince a considerable section of urban population in Colombo, Gampaha and Kalutara Districts for their national and religious causes. The success of both the JHU and the JVP in 2004 election suggests that 'national unity' has become an important concern for the majority Sinhala population.

The chaos generated in selecting the speaker at the thirteenth Parliament session on 22 April 2004 shows the significance of the JHU monks in determining political process in Sri Lanka.[64] While the JHU casted the critical two votes (out of 110 against 109) in electing the former Minister of Justice, Mr W.J.M. Lokubandara of the UNP (Opposition) as the new speaker, the monks of the JHU also faced abuse within the Parliament from the UPFA Government benches and outside the Parliament by unidentified persons often associated with the JVP. When the JHU Member of Parliament, Venerable Athurāliyē Rathana, began to speak in the Parliament congratulating the elected speaker, he was disturbed by the Government peers, particularly by JVP MPs by making noises, calling names as supporters of 'separatists, terrorists and Eelamists', and throwing books at him.[65] Outside the Parliament, an array of offensive posters was posted on walls and billboards accusing the JHU monks for casting votes against people's verdict. This post-election chaos made front-page headline news in the local media.

Post-election events that occurred in relation to the acts of two members of the JHU—Venerable Aparekkē Paññānanda elected from the Gampaha District and Venerable Kathaluwē Rathanaseeha from Colombo District—have created unpleasant reactions in political sphere.

Even before the elections, though Venerable Paññānanda withdrew his nomination, in the election, he won a seat from Gampaha District. Paññānanda had publicly criticized the JHU charging his companions with bribery and corruption. During his campaign, he maintained that the JHU monks accepted black money to finance their pre-election campaign and eventually voiced more support for the UPFA than for the JHU. This event created chaos within the JHU; in the public eye, it made the JHU a divisive political party. The JHU wanted to nominate another candidate on behalf of the withdrawn candidacy of Paññānanda. However, Paññānanda did not want to submit to the JHU's political wish and became a rebellion within the party. Another rebel MP, Venerable Kathaluwē Rathanaseeha, joined the dissident monk Paññānanda to create further chaos within the JHU. Before the Parliament met on 22 April 2004, both of them disappeared mysteriously and UPFA[66] was accused of the abduction.[67] All that happened thereafter is now part of Sri Lanka's very fractured and divided political sphere. Some politicians have been accused of creating this chaos atmosphere within the JHU.

Because of popularity and potential political power of the JHU within the Parliament and chaotic and divisive atmosphere within the JHU, some opponent Members of the Parliament used the opportunity to harass some members of the JHU. The opponents were not even hesitant to abuse fellow parliamentarians of the JHU physically. On 8 June 2004, when Venerable Akmeemana Dayārathana was about to take the oaths going towards the speaker's chair, 'the Government MPs engaged in a struggle to prevent' him doing so by 'grabbing his robe and holding him from his arms'.[68] In this incident two members of the JHU—Akmeemana Dayārathana and Kolonnāvē Sumakgala—were seriously injured due to physical assault and admitted to Fri Jayawardhanapura General hospital.[69] Within a short time, the JHU has become a political party of internal disputes. External pressures also have aggravated this situation. First important key member to leave the party and parliament was Ven. Kolonnave Sumakgala. Soon after Tsunami, on 6 January 2005, Ven. Dhammālōka left the JHU revealing internal divisions. Afterwards, he decided to remain as an independent Member of Parliament. His decision was not well received by the JHU supports. Letters from disheartened Sri Lankans who had actively supported the JHUs success in the election were published in the Internet. Both Ven. Ō. Sōbhita, through his fast unto death in Kandy in June 2005 and Ven. A. Rathana through his political comments on important national issues remain in the public eye as important voices within the JHU. JHU monks were also immensely efficient in helping people who were affected during the Tsunami by providing food, sheltering the victims and building houses.

On 28 May 2004, the JHU MP, Venerable Ōmalpē Sōbhita, published in the *Gazette* a bill entitled *Prohibition of Forcible Conversion of Religion Act* as a Private Member's Bill.[70] The Sri Lankan Government also drafted a bill for the approval of the Cabinet. These two events are meant to fulfil a demand that Sinhala Buddhists made over the last few years with regard to 'unethical conversions' carried out by evangelical Christians in the poor Buddhist and Hindu communities. These bills on 'unethical' conversions bring another phase of religious tensions present in the ethno-religious politics in Sri Lanka. As the youngest and the first monk-led political party, the JHU has already created a significant discourse on its policies and how it will adapt its policies in implementing them in the Parliament and outside it. It has already upset the newly elected ruling party, the UPFA and continues to be an influential factor in Sri Lankan politics.

DISCUSSION QUESTIONS

1. What is the general primary goal of the JHU? What do you think this goal means? What are its members trying to protect?
2. What are the tensions relating to Protestant Christianity in the area?
3. Explain the concept of *dharmarajya* in the JHU.
4. List the controversies within the JHU.
5. What are your feelings about government-controlled NGOs?
6. Summarize the JHU platform.

WOMEN

Buddhist nuns play a vital role in maintaining many of the Buddhist temples and *dharma* centers. These nuns represent a wide spectrum of Buddhist practices. They are well educated, having earned various degrees, even though many migrated from their original birthplaces across Buddhist Asia. Feminist criticism of Buddhist monastic practices is widespread, especially when monks and nuns are compared side by side and throughout various Buddhist practices. The following article analyzes the lives of Buddhist nuns throughout the West and the impact these women have had. It follows the various obstacles that they have had to overcome.

ARTICLE 3
BUDDHIST NUNS: CHANGES AND CHALLENGES

By Karma Lekshe Tsomo

Pioneering Western Buddhist Nuns

Buddhist nuns are a relatively new phenomenon in Western society, appearing only as recently as the late 1960s. Although few in number, nuns in the West have accomplished a great deal in their brief history and have also encountered a number of challenges. Many have achieved high levels of both Buddhist and secular education, maintained high standards of contemplative practice and ethical discipline, helped found and maintain a large number of Buddhist temples and centers, and devoted much energy to serving society. Their greatest challenges have been gaining acceptance for their monastic lifestyle, obtaining material support, obtaining Buddhist education and training, establishing suitable monastic communities, and gender discrimination.

In Asian societies, wearing traditional robes and shaving one's head are expected of Buddhist monastics. The robes and shaved head are symbols of renunciation that elicit respect from the lay community. In Western societies, however, Buddhist nuns and monks are still oddities, and robes and a shaved head may elicit curiosity, admiration, scorn, or abuse. An Asian Buddhist nun is more easily accepted, because Buddhism is viewed as a legitimate aspect of Asian culture. By contrast, a Caucasian Buddhist nun represents a rejection of mainstream religious and cultural values, which may arouse resentment or hostility. The issues are somewhat different for nuns than for monks: nuns are considered odd when they shave their heads; monks are considered odd when they wear skirts. When a nun works to support herself, a shaved head greatly limits employment options. At the same time, the robes and shaved head are valuable because they represent a spiritual commitment and an alternative to consumer culture.

Two successful pioneering Buddhist nuns were the English *bhikṣuṇī* Kechog Palmö (1911–77),[11] who was ordained in the Tibetan tradition in 1966, and the German *bhikṣuṇī* Ayya Khema (1923–97),[12] who was ordained in the Theravāda tradition in 1979. Both of these influential nuns trained extensively in Asia before they began teaching internationally. Other Western *bhikṣuṇīs* who have become successful Buddhist teachers are Pema Chödrön of Gampo Abbey in Nova Scotia; Thubten Chodron of Dharma Friendship Foundation in Seattle;[13] Tenzin Palmo of Dongyu Gatsal

Ling Nunnery in India;[14] Sangye Khandro of Amitabha Buddhist Centre in Singapore; Chi Kwang Sunim of Unmunsa in Korea, now founding a monastery in Australia;[15] and Mu Jin Sunim of Lotus Lantern International Center in Seoul, now teaching in Switzerland.

The Canadian *bhikṣuṇī* Lobsang Chodron (Ann McNeil) is an example of Western Buddhist nuns' dedication. Ordained in 1970, she is the senior nun in the Foundation for the Preservation of the Mahāyāna Tradition, founded in 1974 by Lama Thubten Yeshe and Lama Thubten Zopa. Its monastic organization, the International Mahāyāna Institute, includes 121 nuns of twenty-two nationalities serving in nearly one hundred centers and related organizations around the world.[16] Lobsang Chodron helped start Kopan Monastery in Kathmandu and Chenresig Institute in Australia in the early 1970s, and later Vajrapani Institute and Land of the Medicine Buddha in California. She has taught weekly at Zuru Ling in Vancouver since 1994 and at many other centers around the world. Since 1996, she has also been teaching in the Canadian federal prisons in British Columbia. Chodron is in the process of establishing another monastic center, Kachoe Zung Juk Ling, outside of Vancouver.

Because nuns receive less financial support than monks, their communities often develop in conjunction with established centers. Chenresig Nuns' Community, in Australia, evolved as nuns helped create Chenresig Institute and began to live together as a group. The twenty nuns currently in this community derive many benefits from living near an established center, including the ability to share its facilities and teachers, but they are responsible for their own expenses. Another monastic community of the Tibetan tradition that accommodates nuns is Rashi Gempil Ling, the First Kalmuk Buddhist Temple of New Jersey. The community, which is affiliated with the Asian Classics Institute established in 1993, includes four nuns (two *bhikṣuṇīs* and two *śrāmaṇerikās*). They study with Geshes Lobsang Tharchin and Michael Roach alongside monks and laypeople, many of whom have taken lifelong vows of celibacy. The nuns participate in the twice-monthly confession ceremony on a regular basis and are active in establishing Diamond Mountain, a retreat center in Arizona that will include monasteries for women and men, in addition to schools for children, retirement housing, Buddhist colleges, a translation bureau, and an arts center. Hundreds of other nuns in the West have also taught and served diligently without fanfare for years.

Both in Asian and in Western Buddhist communities, nuns have faced considerable difficulties gaining recognition when they serve as teachers—a traditionally male role. Some still serve in supportive roles as translators, administrators, attendants, and caretakers, leaving the role of teacher to monks and laymen. Many of the challenges they face relate directly to hierarchical gender expectations and assumptions about the subordinate role of nuns that accompany the Buddhist traditions to the West.

Struggles and Successes

The wide variety of Buddhist nuns' experiences can be traced to how they begin their ordained life. Most women who seek ordination have similar motivations—a commitment to study and practice the Dharma. Beyond that, however, their goals and expectations vary considerably. Some are interested in joining a community, while others are attracted to the idea of wandering freely. Some receive training in a monastic environment for years before ordination; others have no interest in monastic protocol. Some seek ordination in a well-structured, well-disciplined community; others seek ordination with the intent of living independently. The diverse paths nuns choose yield a host of dividends and deficits.

In contrast to most Catholic nuns, Buddhist nuns in the West lack established orders to provide material support, monastic training, and spiritual guidance. There is no mistress of novices to train aspirants and no systematic program of monastic formations. Some nuns seem content to live independently. Others imagine they are entering a large, supportive community and are rudely awakened to find themselves completely on their own. Without any sense of monastic unity, no guides to train or advise them, and little understanding on the part of the laypeople around them, these nuns grapple with feelings of isolation, abandonment, and rejection.

The Sangha, a broad community of Buddhist monastics spread throughout the world, offers the ordained membership in a broader community. This broad membership does not satisfy the basic needs of a newly ordained nun for food, shelter, and training. It is not uncommon for a Western woman to become ordained as a nun only to discover that the master who ordained her has flown off to his next destination. The fresh novice is left with nothing to eat, nowhere to stay, no training, no arrangements for further education, and sometimes without any knowledge of the precepts she has pledged to uphold for life or how to wear the robes properly. After the first flush of excitement about ordination subsides, she begins to realize her aloneness and lack of a supportive community.[17] As Thubten Chodron relates, "Few monasteries exist in the West and, if we want to stay in one, we generally have to pay to do so because the community has no money. This presents some challenges: how does someone with monastic precepts, which include wearing robes, shaving one's head, not handling money, and not doing business, earn money?"[18]

It is widely assumed that Western Buddhist nuns live in supportive communities and are financially supported by monastic orders. On the contrary, most are on their own when it comes to obtaining food, shelter, education, training, health care, and spiritual guidance. Unlike most Christian nuns' communities, the center where a Western Buddhist nun works may or may not provide her with room and board. Even in Asia, where monks are well respected and cared for, nuns are not always accepted as Sangha and do not receive the same level of material support as monks. In the West, traditional systems of providing monastics with the necessities of life are not yet in place and Western laypeople do not express much interest in developing them. Despite egalitarian ideals and increasing feminist awareness, most nuns are left to fend for themselves, though Western monks generally find sponsorship. Laypeople may even ask nuns, "Why should nuns deserve respect and support?" They may say, "My practice is just as good as yours. Why do you think you are better than I am?" These types of comments reveal a serious lack of understanding of the purpose and principles of monastic life. Often, with Roman Catholicism as the only available model, Western Buddhists themselves are unclear about what ordination means.

Monastic training in Buddhist cultures is traditionally provided by individual monasteries, yet it usually much less formal than the Catholic model. In addition, monasteries for women are generally few in number and poorly supported. The lower status of women in Buddhist societies is reflected in the lower status of nuns. Because nuns often have limited access to Buddhist education, training, and material support, they are less likely to emerge as teachers and role models for the next generation. This self-perpetuating cycle is beginning to change as nuns join forces to encourage each other and improve their circumstances, as has occurred rather dramatically in Korea, Taiwan, and Vietnam.[19]

Monasticism in the West: An Open Dialogue

As Buddhism has become more popular in the West, both practitioners and outside observers have begun to notice the lack of female monastics and teachers and the lack of equal opportunities for education and ordination for women. Recognition of the neglect and inequalities Buddhist women have historically experienced has precipitated dialogue among nuns of various Buddhist traditions. This growing awareness has motivated scholars and practitioners to conduct research into Buddhist women's history and current circumstances, particularly the history of the *bhikṣuṇī* lineage and the prospects for instituting or restoring the *bhikṣuṇī* ordination in the various Buddhist traditions. One important achievement has been increased support for Buddhist women's projects in Asia. Another result has been an increased awareness of the need for viable Buddhist monastic institutions for women in the West.

The aim of creating monasteries in the West is not simply to duplicate Asian monastic structures, but to develop a monasticism appropriate for the West. Monastic practice is an attractive option for only a few Western Buddhists at present; far greater numbers prefer lay practice and seek lay teachers, who they think will be better able to understand their problems. Due in part to stereotypes of Christian monasticism, and in part to the tension between the monastic ideal and the secular work ethic, many Western Buddhists are not yet convinced that monastic practice is a meaningful option. Others who agree with the traditional view that monastic institutions are necessary for Buddhism to take root on Western soil are concerned about how monasticism can be profitably adapted in the West.

Another facet of this dialogue is the recognition that women attracted to Buddhism are not always attracted to monastic life. Until suitable monastic environments for women evolve, nuns must live in lay communities, join mixed communities (comprised of monks, nuns, and lay practitioners), train in Asia, or manage on their own. Each option has advantages and limitations. Living in a mixed community of nuns, monks, and laypeople may offer social and educational advantages, but is a very different experience than living in a supportive community of nuns. Training in Asia may be culturally enriching, but may also entail foreign-language acquisition, health risks, visa restrictions, and poverty. Striking out on one's own may offer an independent lifestyle, but provides no systematic study program, training, sense of community, or psychological support. Some Western nuns with independent means are itinerant, taking teachings and doing retreats around the world. Some study with a traditional teacher in Asia for a period of time, but find no suitable community or support when they return to their home countries. Those who find it difficult to be monastic without a monastery generally leave the robes altogether.

The importance of finding a well-qualified teacher cannot be overstated. In traditions that lack qualified female teachers, nuns must rely upon male teachers. As long as both teacher and student carefully observe the precepts, the relationship may be highly beneficial. If open dialogue and gender equality are lacking, however, the relationship may replicate patriarchal patterns and become dysfunctional. Asian nuns who have been raised to accept gender disparities may be willing to accept subservient roles and inequities in the teacher/student relationship, and may even consider them character-building, but few young women in the West are attracted to traditional hierarchical models of monastic life. Those few who do become nuns generally seek a situation with educational opportunities and a measure of independence.

The disparity between the ideal of sentient beings' equal potential for enlightenment and the social reality of gender discrimination can be very discouraging, especially for new nuns raised in the West. Nuns generally play important roles in community organizing, counseling, administration, and communications, but rarely receive training, recognition, or due compensation for their skills. They often encounter biases with regard to their gender and their celibate status. Anxious to implement the Buddhist ideals of compassion and selflessness, they may disregard discrimination or attempt to resolve frustrations by applying appropriate Dharma teachings. Receiving little psychological support or encouragement, nuns in a Western Buddhist environment face high expectations and incessant demands on their time.

When Buddhist institutional structures neglect nuns, whether in Asia or in the West, they have fewer chances to emerge as teachers and role models, and the cycle of subordination becomes self-perpetuating. Without opportunities for systematic education and long-term retreats, the traditional patriarchal assumptions about women's inferiority continue to thwart nuns' advancement. Teachers may affirm the spiritual equality of women, but without gender equity in practice these assurances ring hollow, and may even mask discrimination against women. At international conferences, it is common to hear affirmations of women's equality, but rare to find practical proposals to remove blatant inequalities. Ironically, even in the much-heralded woman-friendly West, nuns may find themselves ignored or rejected until they prove their worth as teachers, translators, authors, fund-raisers, or administrators. By the time a nun has become successful, she has already gained confidence and self-reliance, and the support system she so desperately longed for as a new nun is no longer needed.

Even though many Western nuns are well educated and skilled, they may find themselves marginalized, both in Asian Buddhist environments, due to different languages and customs, and in Western Buddhist environments, where lay Buddhist practice is emphasized. To make matters worse, male teachers may regard nuns who are knowledgeable about Buddhism as a challenge to their authority. Eventually Western nuns find they must balance the benefits of Buddhist practice against their disadvantaged gender status. The eight special rules (*gurudharmas*) that subordinate nuns to monks were clearly designed to prevent nuns from challenging monks' authority in Buddhist institutions. If it can be shown that the eight *guru-dharmas* were not spoken by the Buddha, as recent research indicates, this would effectively upset the traditional apple cart and eliminate barriers to women's full and equal participation in Buddhist institutions.[20]

Paradoxically, despite the stark inequalities in opportunities and support, nuns seem to have a much easier time maintaining celibacy. Contrary to the popular myth that women are sexually uncontrollable, the dropout rate among nuns is much lower than among monks. Usually the decision to leave monastic life is more a product of frustration with the lack of a conducive environment for personal growth than the desire for a sexual relationship. Because of their staying power, nuns are likely to exert a strong influence on Buddhist institutions far into the future.

Revisiting the Ordination Issue

Western nuns have been leaders in the movement to reinstate full ordination for Buddhist women in traditions where it is not currently available.[21] Because the struggle for equal rights for women has a long history in Western societies, Western nuns are generally more concerned about achieving equal rights within Buddhism than Asian nuns, who either have already achieved a semblance of equality with monks (as in Taiwan and Korea) or do not regard gender equality as an appropriate pursuit for Dharma practitioners (as among many Thai and Tibetan nuns, for example). Proponents of full ordination for women point to the vitality and effectiveness of the *bhikṣuṇī* Sangha in Taiwan and Korea as evidence of the benefits fully ordained nuns can bring to Buddhism. Equally important, the movement to reinstate full ordination is a historical step in intercultural exchange among nuns.

Western nuns generally find *bhikṣuṇī* ordination extremely beneficial for many reasons. First, they receive the karmic benefit in maintaining the full complement of precepts. Second, they appreciate the practical benefits of having a time-tested code to guide their personal conduct and as an aid in developing mindfulness. Third, they gain confidence and benefit psychologically from becoming full-fledged members of the Buddhist Sangha, rather than being relegated to novice status indefinitely. Fourth, whether they receive ordination in Taiwan, Korea, or the West, they gain inspiration from belonging to a community of Buddhist nuns who model the ideals of discipline, dedication, spiritual confidence, and loving-kindness. Finally, as full members of the Sangha community, they become *fields of merit* and therefore worthy of support.

So far, gaining opportunities for full ordination has primarily been a concern of a core of educated Asian laywomen and Western Buddhist nuns. Efforts to educate a larger number of Asian nuns and monks about the issues will require a team of multilingual activists with a strong commitment to these issues. Many Asian nuns are too humble, too poor, or too educationally disadvantaged to take an active interest in receiving full ordination. Some also have been discouraged from seeking full ordination by their teachers, who advise them that novice precepts are sufficient for nuns. By contrast, most Western Buddhist nuns who have received *bhikṣuṇī* ordination feel that all nuns who wish to take full ordination should have the opportunity. These nuns have moved the issue forward simply by taking the ordination themselves and discussing the issue of full ordination wherever they go. This strategy of active commitment and education is designed to gain the support of those who oppose the full ordination of women or are simply unfamiliar with the issues. These efforts have had a number of unforeseen positive results. The full ordination issue has helped nuns find their voice and stimulated research on the *bhikṣuṇī Vinaya*. It has facilitated dialogue, not only among nuns of various traditions, but also among nuns, *bhikṣu* scholars, and male-dominated Buddhist institutions.[22] A tremendous amount of research, community building, and dialogue remains to be done, but awareness of gender-related issues has increased and will surely continue.

Many monks and nuns in the Tibetan and Theravādin traditions are unaware of the large, vital *bhikṣuṇī* Sanghas that exist in Taiwan, Korea, and Vietnam, but anyone who travels to these countries cannot fail to be impressed with the erudition and hard work of these nuns and the benefits they bring to the Dharma and society. Coming into contact with so many well-educated, well-disciplined nuns inevitably causes monks to reevaluate the situations of nuns in their own countries and cultures. Contact with vital *bhikṣuṇī* communities can have the effect of generating a true concern for and commitment to improving conditions for nuns, but it can also be somewhat intimidating. Seeing the excellent conduct and social action programs of nuns in Taiwan, for example, can cause monks to become painfully aware of the inadequacies in their own traditions, not only among the nuns but also among the monks. There is an obvious need for improved monastic training in some traditions, but changes require stable monastic communities and sincere conviction about the benefits of monastic discipline.

Another reason for supporting full ordination for Buddhist nuns is strategic. To ensure the success of Dharma in the West, equality in the Sangha is essential. If Buddhism is to be taken seriously in the long term, women must be granted equal opportunities in Buddhist institutions. Therefore, it is of great historical significance that full ordination for women be made available in all Buddhist traditions, and the sooner the better. Asian monks who lack a *bhikṣuṇī* Sangha in their own traditions hesitate to accept as valid the *bhikṣuṇī* lineages of other traditions for several reasons: (1) doubts about the validity of the extant *bhikṣuṇī* lineages (that is, their purity and unbroken continuity); (2) doubts about the authenticity of the ordination procedures; and (3) doubts about the acceptability of conducting future *bhikṣuṇī* ordinations using *bhikṣus* and *bhikṣuṇīs* of two different *Vinaya* lineages. The prescribed procedures for conducting a dual ordination with ten *bhikṣus* and ten *bhikṣuṇīs* appear to have lapsed periodically in China, Korea, and Vietnam, but there is no evidence that the *bhikṣuṇī* lineage has been interrupted at any time in Chinese history. There are differences of opinion about whether a *bhikṣuṇī* ordination con ducted by *bhikṣus* alone is valid or not.

Where *bhikṣuṇīs* do not exist to ordain *śrāmaṇerikās*, novices are selected and ordained by monks, who may not carefully assess their qualifications and may not assume responsibility for their care after ordination. The selection of nuns by monks has resulted in many difficulties, even tragedies, and caused some nuns to recognize the importance of women's participation in the ordination process. *Bhikṣuṇīs* must learn to conduct the *śrāmaṇerikā, śikṣamāṇā,* and *bhikṣuṇī* ordinations and assume active responsibility for the training of future generations of Buddhist nuns. Now that sufficient numbers of Western nuns have received *bhikṣuṇīs* ordination and have the requisite twelve years of seniority, this has become a possibility. If Western *bhikṣuṇīs* learn to conduct the ordinations for nuns and are willing to serve as mentors, they will be able to ensure that the women who join their order are carefully selected and properly trained. This requires not only that these *bhikṣuṇīs* themselves are thoroughly trained and educated, but also that they are prepared to train and support the nuns they ordain. Monastic communities as conducive environments for the education and training of nuns—and monks—are essential for the authentic transmission of Dharma to the West.

Women, Monasticism, and International Dialogue

Conferences, seminars, and informal gatherings held around the world in recent years have stimulated research into women's past and present roles in Buddhism and have helped envision an expansion of these roles. Most of these gatherings have been organized informally as a result of private initiatives, without the support of traditional Buddhist institutions. The objectives have been women's empowerment, networking, meditation training, and education on issues relevant to women. In the course of these discussions, the need for an organized, educated, unified, self-sufficient order of nuns has been clearly recognized, but much work remains to make it a reality.

The first international meeting of Buddhist nuns in recorded history occurred in Bodhgaya, India, in 1987. Although the impetus for this first conference was specifically to address the pressing issues facing Buddhist nuns at that time, the consensus at the close of the conference led participants to forge a coalition among nuns and laywomen called Sakyadhita, the first international Buddhist women's organization. Sakyadhita's major objectives include improving education, communications, facilities, and ordination opportunities for Buddhist nuns, and considerable progress has been made on all counts. This organization continues to foster cooperation among nuns and laywomen around the world through local, national, and international gatherings, as well as through publications and other projects.

Nuns have actively promoted dialogue among Western monastics in general. Since 1995, Tenzin Kacho and Thubten Chodron have helped organize an annual Western Monastic Buddhist Conference. Dialogue has centered on such varied topics as friendship, lay/monastic relations, interpersonal communications skills, leadership, counseling, teaching skills, and prison work. Participants sense that ordained teachers have a special role to play in maintaining the authenticity of the Buddhist traditions, whereas lay teachers are more interested in adapting those traditions. If this is the case, then continued dialogue between monastic and lay practitioners is essential for achieving a balance between authenticity and adaptation.

In 1996, a three-week educational program, called "Life as a Western Buddhist Nun," was held in Bodhgaya, attended by eighty nuns (including twenty from Tibet and the Himalayan region) and twenty interested laywomen, several of whom subsequently received ordination. In a message sent to participants in this program, H.H. Dalai Lama observed that, "In the past, in many Buddhist countries, nuns did not have the same educational opportunities as monks, nor access to the same facilities. Due to prevailing social attitudes nuns were often treated or regarded in ways that are no longer acceptable today. … It is heartening to observe … that Buddhist women are casting off traditional and outmoded restraints." The talks from this program have been compiled by *bhikṣuṇī* Thubten Chodron in *Blossoms of the Dharma: Living as a Buddhist Nun*.[23]

Several conferences on the *bhikṣuṇī* issue have been held in Dharamsala, *bhikṣuṇīs* scholars have not been invited to participate in them. Despite the supportive attitude of H.H. Dalai Lama on the matter of higher ordination for women, these discussions clearly indicate that many *bhikṣu* scholars and monastic administrators oppose or are noncommittal on the subject. The issue is complicated by language difficulties, the lack of historical documents to verify the unbroken continuity of the Chinese *bhikṣuṇī* lineage, age-old attitudes of gender discrimination, and politics, both religious and secular. Hopefully Buddhist institutions will begin to recognize that equitable dialogue on the *bhikṣuṇīs* issue must involve not only *bhikṣuṇīs* scholars but also those most concerned: the *bhikṣuṇīs* themselves.

DISCUSSION QUESTIONS

1. Name at least two pioneering nuns, and list their accomplishments.
2. List three common struggles for Buddhist nuns.
3. Describe a monastic practice that the article identifies as "Western-appropriate."
4. Discuss efforts toward full ordination.
5. When and where was the first international meeting of Buddhist nuns held?
6. What did the Dalai Lama observe at the "Life as a Western Buddhist Nun" conference?

JAINISM

INTRODUCTION

One who neglects or disregards the existence of earth, air, fire, water, and vegetation disregards his own existence, which is entwined with them.—Mahavira

Jains form a very small percentage of the Indian population, yet this religion impacts the political, social, and economic landscape of India in a disproportionately large way. Jainism developed almost alongside Buddhism. Some claim that Jain probably developed first, but either way the two religions were reacting to similar strife. Both Jainism and Buddhism eventually developed teachings that reject the authority of the *Vedas* and lay out paths to reach liberation from the cycle of *samsara* though rebirth and *karma*. The founder of Jainism is thought to be a man named **Vardhamana Jnatrputra**, and although this belief is contested, many scholars place him as a contemporary of the Buddha. Vardhamana's story is also very similar to the story of Siddhartha. After the death of his parents, when he was around the age of 30, he renounced his life as a prince inside the warrior caste and roamed around India as an ascetic monk. After he spent 12 years practicing extreme forms of asceticism, his followers named him **Mahavira**, or "Great Hero." He believed he was the 24th—and the last—in a long line of *jinas*, or "conquerors of world desire."

In Jainism, the 23 predecessors of Vardhamana are known as ***tirthankaras***, a term that means "bridge builder." *Tirthankaras* conquered *samsara* and have effectively built a bridge from this life to salvation. Their lives were filled with teaching, and their teachings focused especially on how to reach salvation. Vardhamana used both the words *moksha* and *nirvana* to describe "salvation."

Although no historical evidence exists for the first 22 *tirthankaras*, followers of the 24 *tirthankaras* are known as *Jainas* or *Jains*. Jainism may recognize gods, but it does not rely on them. However, there is definitely no creator god. In fact, the world is understood as always existing, or eternal. The *tirthankara* named Mahavira taught that the world was never created and will never be destroyed. Like a circle, is the world has no beginning or end. Jainism includes the philosophical thought of dualism. Everything is either of *jiva* ("soul") or *ajiva* ("matter"). These concepts are also eternal and have always existed.

The *jiva*, or soul, is what reincarnates, similar to that in Hinduism. The soul experiences the consequences of *karmic* action and can become liberated. In order to attain salvation, the soul that is tainted with *karma* needs to get rid of it. Like Mahavira, Jains can practice fasting, meditation, penance, or yoga, or they can study and recite the scriptures. For some Jains, the ultimate ascetic observance is **sallekhana**, or "fasting to death." Ridding the body of *karma* restores or purifies the soul. Reaching salvation would require stopping the flow of new *karma* and then eliminating the old. Similar to teachings in *Theravada* Buddhism or the stress of the *Upanishads*, in Jainism the path to salvation relies on individual effort. To achieve this salvation a Jain could follow the **Five Great Vows**.

The first Vow includes **ahimsa**, or "nonviolence." In striving to prevent violence of any kind, Jains are vegetarians. Some followers limit their intake of dairy products, avoid root vegetables, or avoid eating after sunset. However, *ahimsa* is not limited only to physical violence; there is also *ahimsa* of the mind and of speech. A follower needs to have right thought and right speech, which include engaging in kind and compassionate language. Empathy should result from seeing oneself and the world properly. The other four Vows are (1) to speak truth; (2) not to steal or take what is not given; (3) celibacy for monks, or faithfulness in marriage; and (4) nonattachment to material things. Ultimately, these four Vows are also connected to avoiding violence and accumulating *karma*.

Forgiveness is also a key concept within the religion. Jains embark upon an eight-day spiritual journey of reflection and repentance, similar to the intentions of *Hajj* in Islam, as they begin celebrating one of the most important yearly festivals in Jainism: **Paijusana**. During these eight days, Jains emphasize forgiveness and self-purification. It is a time for them to address their *karmic* actions and repent for any wrongs they've committed. The festival ends with **Samvatsari**, or the "Day of Forgiveness," when followers receive absolution to start the New Year with a clean slate. Jainism also exalts principles of equality and interdependence. Mahavira taught that the Earth, and everything in it, is bound with the existence of humans, and to disregard one essentially means disregarding the other. Every living life-form should support all other life-forms. Humans need to support the plants, pebbles, insects, animals, etc., as these life-forms support humans. Some Jains go to extreme lengths to avoid harming any kind of life-form.

As a result of a disagreement over the true meaning of Jainism, since 80 BCE Jain history has seen two prominent monastic sects. The first sect is the **Svetambara**, meaning "white-clad." Rejecting the idea that monks must be nude, these devotees wear long white cloths. They are found mostly in northern India and are considered liberal in their interpretations. The second sect is known as the **Digambara**, meaning "sky-clad" or "naked." These devotees are mainly from southern India and are believed to be more conservative in their interpretations. Women cannot enter monasteries or temples, for example.

ECOLOGY

Jainism is often spoken of as a religion that is full of ecological harmony—perhaps because so many Jain concepts translate easily into great ecological practices. Whether these practices have actually materialized is a question for another article. However, nonviolence, self-restraint, avoidance of wastefulness, kindness to animals, and vegetarianism are some of the concepts that contribute to this ecological harmony. The following article discusses some of these principles and how they can have a positive impact on environmentalism through ecological theory and ethic.

ARTICLE 1

"THEY'LL KNOW WE ARE PROCESS THINKERS BY OUR…" FINDING THE ECOLOGICAL ETHIC OF WHITEHEAD THROUGH THE LENS OF JAINISM AND ECOFEMINIST CARE

By Brianne Donaldson

A Solution: Syādvāda and Mutual Immanence

Both Jainism and process thought, because of their perspective on relational relativity, can address the ecofeminist critiques of metaphysics discussed above. Already we see that Whitehead's work regarding subjectivity and novelty navigates ecofeminist hurdles insofar as real subjective creativity within individual processes bridges dualistic separation and hierarchy. There is no "separative self" to speak of that exists independent of particular events and networks of continuous mutual response. Rather, there is an embedded "debt toward that which gives and renews life," a give-and-take in the midst of relational multiplicity.[1]

Further, the locus of transcendence changes so that neither matter nor consciousness are best defined by what is above or ahead—something one is *progressing* toward—but rather by the *process* itself, where the spark of creative agency flashes *within* every novel moment. In fact, similar interpretations of transcendence have already influenced ecofeminists such as Ignacio Ellacuria's *intracosmic model of transcendence*—influential for later liberation and ecofeminist work—in which the "future emerges from within the matrices of relations."[2] Yet, I think, there is even more we can glean from Whitehead on this point of transcendence by looking to his work on mutual immanence.

In *Adventures of Ideas*, Whitehead describes mutual immanence as centrally descriptive of his metaphysics, and indeed of the entire universe as process. Whitehead draws upon the common ground shared by Aristotle, Plato, Epicurus, Lucretius, and Leibnitz when he posits "the diverse notions of communication between real individuals" and "the diverse notions of the mediating basis in virtue of which such communication is attained."[3] For Whitehead, if anything could be generalizable or transcendent in the universe, it is the reality of mutual immanence as "a natural matrix for all things," or a "medium of intercommunication" that unifies everything in the natural world.[4]

Ecofeminists, however, might claim that this is merely another universal, ultimately capturing everything with a transcendent category or over-arching principle. But in Whitehead's formulation, "there is no universality transcending the mutual immanence of all actualities, which already harbour possibilities, categories, and principles. It is a non-category, a non-principle. It is not a unity, but pure difference. It is not form, but only connection,"[5] where "universality defined by 'communication' can suffice."[6]

This communication results in linguistic dialectics within Whitehead, most notably the God/World relational dialectic that he constructs in *Process and Reality*. Here he sets out his antithetical proofs embodying the phenomenon of mutual immanence:

> It is as true to say that God is permanent and the World fluent, as that the world is permanent and God is fluent. It is as true to say the God is one and the World many, as that the World is one and God many … It is as true to say that God transcends the World, as that the World transcends God. It is as true to say that God creates the World, as that the World creates God.[7]

The mutual immanence expressed in these proofs finds equivalent articulation in Jainism as *syādvāda*, or dialectic of relativity. Translatable literally as the "maybe doctrine," or more accurately as the "doctrine of conditional or qualified assertion," *syādvāda* in a non-exclusivist, non-absolutist form of speech, rooted in a metaphysics of relativity.[8] As described by Bimal Krishna Matilal, "'syat' means, in the Jain use, a conditional YES. It is like saying, 'in a certain sense, yes.'"[9] As a participle meant to convey indefiniteness, Jains use it paradoxically, according to Long, to disambiguate language or "to coordinate the exclusive, one-sided claims made by various competing schools of thought with partially valid perspectives, or nayas."[10] The seven applications of *syādvāda* (a combination of *syat* or "in some respect" with *eva* or "absolutistic import") results in seven possible truth claims that state:

1. In a certain sense, x exists.
2. In a certain sense, x does not exist.
3. In a certain sense, x exists and does not exist.
4. In a certain sense, x is inexpressible.
5. In a certain sense, x both exists and is inexpressible.
6. In a certain sense, x does exist and is inexpressible.
7. In a certain sense, x does not exist and is inexpressible.[11]

Matilal suggests that Jains use this sevenfold formula as a method of refined concession within philosophical debate:

> It concedes the opponent's thesis in order to blunt the sharpness of his attack and disagreement, and at the same time it is calculated to persuade the opponent to see another point of view or carefully consider the other side of the case. Thus, the Jain use of "syat" has both; it has a disarming effect and contains (implicitly) a persuasive force.[12]

The persuasive force embodied in the Jain employment of *syādvāda* as linguistic destabilization is similar to the paradoxical transcending principle in Whitehead's metaphysics which takes the form of mutual immanence.[13] As with *syādvāda*—the function of which is to "demonstrate the incompleteness, the partiality, of the truths expressed in non-Jain perspectives," Whitehead's idea of mutual immanence acts as a similar dialectic where, "universality and relativity, singularity and relationality, creativity and extension are differently related: they are *manifolds in mutual immanence* of which time, space, ideality (eternal objects), extension, and creativity are expressions of their mutual, universal incompleteness."[14]

Far from the hierarchy of height and progress implied in standard notions of transcendence that rightly earn ecofeminist critiques, the concepts of *syādvāda* and mutual immanence are as much a recognition of inextricable relationality as they are intellectual tools to bring to ethical contexts. Faber's interpretation of mutual immanence reinforces the point clearly, saying, "[mutual immanence] is a *critical* notion that, in refuting any transcendence of categories and principles, denies anything the status of origin, source, ground, aim or goal beyond the nexus of happenings itself. It is anti-hierarchical!"[15] It is from this point, having confronted ecofeminist concerns over transcendence as well as universalization, that we can begin to synthesize an extension of Whitehead's formula with the Jain notion of *ahimsa* and an ecofeminist ethos of care.

Whitehead's Ecological Ethics?

While Whitehead's metaphysics appears to address ecofeminist concerns over dominant western formulations of transcendence and universalization, the question of ethics is not fully answered. We are still left with the practical inquiry, "How do we live?" Although Whitehead recognizes ethics as one of several influences within the becoming process of the universe, he more often describes measures of beauty and intensity as the ultimate goal of processes.[16] Yet, this does not mean that Whitehead has nothing concrete to offer about how we should get on in the world.

If Whitehead's metaphysics have shown us anything thus far it is that any ethic has to be consistent with the relativity of misplaced concreteness and the dialectic relationality of mutual immanence. Once again we can look to the Jains for direction. L. M. Singhvi's "Jain Declaration on Nature" reads:

> Because it is rooted in the doctrine(s) of *anekāntavāda* and *syādvāda*, Jainism does not look upon the universe from an anthropocentric, ethnocentric or egocentric viewpoint. It takes into account the viewpoints of other species, other communities and nations and other human beings.[17]

Where the two doctrines of *anekāntavāda* and *syādvāda* begin to concretely affect action is in their location within the larger metaphysical context of Jainism. The central principle in Jainism, entailed by its overall cosmology, is *ahimsa*, often translated as "nonviolence," or more accurately "the absence of even the desire to do harm."[18] Ethically, according to Long, this is the principle which can give the Jain doctrines of relativity (and by extension, Whitehead's misplaced concreteness and mutual immanence) their "critical edge."[19]

Within the Jain *darsana*—a notion modeled after the Greek "philosophia" that denotes a total worldview and way of life—Jainism, similar to the process worldview, sees itself "as co-extensive with the nature of reality itself—with the true nature of things."[20] The Jain *darsana* describes an internally coherent system that assumes a logical relationship between the ethical principle of *ahimsa* and the dialectical relativity of *anekāntavāda* and *syādvāda*. In Jainism, the cardinal sin is, after all, one-sided interpretation.

Long uses the example of Nazism as a test case in which truth claims against others (i.e., that some groups of Jews, Gypsies, etc., are fundamentally different from the Nazi) must be conditioned by its contrary (those people who are different from the Nazi are also like him or her). A simultaneous recognition of the difference and similarity "would complete [the Nazi's] one-sided perspective, thereby logically negating the violence inherent in it."[21] In this way, the dialectics of relativity function as a philosophy of nonviolence. Writes Long:

> If one can presume, a priori, that all claims which could entail injury to others must necessarily contain a one-sided affirmation—which will, in all likelihood, typically be

an affirmation of some seemingly unbridgeable distance between the speaker and the object of his intended violence—and then correct this one-sidedness with an affirmation of its contrary—namely, the common humanity, or "beingness," of the speaker and the object of his intended violence—then one should be able to avoid the problem of one's pluralistic interpretive method inadvertently justifying truth-claims which advocate or approve of violent acts.[22]

In a similar way, Whitehead's mutual immanence and misplaced concreteness posit metaphysical perspectives of incompletion that, when complemented by mutual requirement and inherent relationality, negate the one-dimensional otherness necessary to justify violence. Further, in non-anthropocentric metaphysical traditions, like Hinduism, Buddhism, and Jainism (and arguably now, process thought as well), one is:

> Enjoined to follow something like the Golden Rule because all beings—and not only humans—ultimately constitute one Being, one Oversoul—or *Brahman* ... or [as in the theistic traditions of the West], all beings are affirmed to be children of God ... [or as in Buddhism] all beings are dependently co-originated and ultimately no different from one "self".[23]

Process philosophers are not of one mind regarding violence—more aptly called "violation" by Keller, to signify uniquely human systems of power and repetitive brutality[24]—and nonviolence. But, while Whitehead himself contends that "life is robbery,"[25] the central affirmation of process thought—that experience is shared by all entities—offers not only a logical basis to avoid complicity with as much violence as possible, but also a "positive injunction to work for the good of all beings."[26]

For Whitehead, this positive injunction *beyond* non-violation might look very much like the Jain *darsana* rooted in this embodied *way of ahimsa*. It might look very much like an ecofeminist philosophy that, according to Kheel, "is not so much an ethic as a consciousness or ethos":

> It is a "way of life" or a mode of consciousness that invites us to be "responsible," not in the sense of conforming to obligations and rights, but in the literal sense of developing the ability for response. It is an invitation to dissolve the dualistic thinking that separates reason from emotion, the conscious from unconscious, the "domestic" from

the "wild," and animal advocacy from nature ethics. It welcomes the larger scientific stories of evolutionary and ecological processes, but never loses sight of the individual beings who exist within these larger narratives.[27]

Surprisingly, we find a strikingly similar definition of "peace" in Whitehead's *Adventures of Ideas*. As one of the five qualities of civilization, Whitehead puts the ethos of peace this way, "[h]ere by the last quality of Peace, I am not referring to political relations. I mean a quality of mind steady in its reliance that fine action is treasured in the nature of things."[28]

With a metaphysical system that describes the "nature of things" as relative in their particular concreteness to one another and in mutually immanent relations of co-dependence and creative novelty, Whitehead's ethic is not an absolute, totalizing demand. Just as his metaphysics is rooted in paradox, so too is any process ethics. Like the reality in which ethical action resides, Whitehead might well consider "fine action" to be that ethos of genuine responsiveness that develops out of recognizing the subjective value and interdependence—and the similarities and differences therein—of all matter and all beings, human or non-human.

Like the relationship in Jainism between the doctrines of *anekāntavāda* and *syādvāda* and the core affirmation of *ahimsa* as a lived expression of metaphysical claims, a "care-sensitive" ethics in process thought honors the multiplicity of priorities within human care and concern in a given moment, even as it honors the metaphysical foundations of relationality and immanence that characterize our very existence.

This contextualized understanding of care and responsiveness is lived out ecologically in the Jain code of conduct that stresses nonviolence in thought and deed, gentleness toward plants and animals, comprehensive vegetarianism, self-restraint, avoidance of waste, and charity.[29] It is lived out in the ecofeminist perspective that includes in its ecological evaluation the systematized violation of non-human animals in food production, clothing, entertainment, and captivity, and that expands ethical exploration to social constructions such as gender identity that contribute to the abuse of animals and our shared ecological support systems.[30]

Many process thinkers have already begun to explore the rich possibilities of Whitehead's metaphysics for addressing ecological violence.[31] There is, however, still ample room for elucidating and encouraging concrete actions in our particular contexts; actions that, per Whitehead's poetic description "[dwell] upon the tender elements in the world, which slowly and in quietness operate by love."[32] Our ability to dress in such vestments of boundary-crossing care will depend on a willingness to embrace the enchanting irrationality of Whitehead's system; a system whose ecological way of being compels us beyond Scholtes' unifying vision of group identity toward disorderly collisions with the frequently invisible people, animals, or ideas that are entwined with our daily existence. It is precisely when these differences are allowed to surface and bump into one another that we need a metaphysics spacious enough to illuminate the possibilities for ethical action.

Whitehead's metaphysics implies that our ecosystems, neighbors, and animal cousins are capable of receiving and contributing to our responsiveness and care, when we offer it toward those beings who, and places which, our basic daily choices such as eating, speaking, and spending affect. In those metaphysically mundane moments with fork, flyswatter, or dollar bill in hand, the possibilities we affirm and refuse allow us to take seriously the "unfading importance of our immediate actions."[33]

Unlike Scholtes' tune, where love cannot remove the distance between the "they" and "we" named in the title, Whitehead recognizes the folly and power required to maintain such illusory, stabilizing distinctions. Love is, after all, nothing if not another boundary dissolved, however disconcerting it may be to relinquish our concentric circles of identity. So let them know we are process thinkers by our quivering chins, our clammy palms, our furrowed brows, determined but not cocksure, compelled toward unnerving encounters with the others who make and unmake us in the not-so-commonplace crossings of implausible love.

Notes

1. Ibid., p. 129.
2. Ibid., p. 44.
3. Whitehead, *Adventures*, p. 135.
4. Ibid., p. 134.
5. Faber, p. 24.
6. Whitehead, *Process*, p. 4.
7. Ibid., p. 348.
8. Long, "Plurality," p. 259.
9. Matilal, p. 52.
10. Long, "Plurality," p. 260.
11. Matilal, p. 55.
12. Ibid., p. 52.

13. Faber, p. 13.
14. Ibid, original emphasis.
15. Ibid., p. 24., original emphasis
16. Whitehead, *Process*, p. 189.
17. L. M. Singhvi, "Appendix: The Jain Declaration on Nature," in C. K. Chapple (ed.), *Jainism and Ecology: Nonviolence in the Web of Life*, Cambridge, MA: Harvard University Press, 2002, p. 220.
18. Long, "Plurality," p. 32.
19. Ibid.
20. Ibid., p. 220.
21. Ibid., p. 34.
22. Ibid.
23. Ibid., p. 35.
24. C. Keller, "The Mystery of the Insoluble Evil: Violence and Evil in Marjorie Suchocki," in J. Bracken (ed.), *World Without End: Christian Eschatology From A Process Perspective*, Grand Rapids, MI: Eerdmanns, 2005, p. 52.
25. Whitehead, *Process,* p. 105.
26. Long, "Plurality," p. 37.
27. Kheel, p. 251.
28. Whitehead, *Adventures*, p. 274.
29. Singhvi, pp. 222–4.
30. Kheel, pp. 250–1.
31. Whitehead, *Process*, p. 343.
32. Ibid., p. 351.
33. Whitehead, *Process and Reality*, p. 351

DISCUSSION QUESTIONS

1. Explain the concept of mutual evidence.
2. Describe the concept of the "maybe doctrine." Explain the logic behind the seven possible truth claims. What do you think about these claims?
3. Why is *syat*?
4. What does the article say about the Declaration of Nature? What do these various viewpoints mean for ecological practice?
5. What is *ahimsa*?
6. What is the concept of *darsana*?

POLITICS

In 2015, the Rajasthan High Court ruled that the Jain ritual of fasting to death is a form of suicide. This effectively made the Jain ritual illegal. The Jains from Rajasthan immediately protested against the ban. This ritual has been an ageless practice within their religion, and Jains do not consider it to be a form of suicide. The Supreme Court of India, the country's highest court, has temporarily lifted the ban until the case is heard before them. The final ruling may not be handed down for several more years. Is this religious ritual, in effect, a form of suicide? The following article breaks down the different forms of suicide and examines how the ritual of fasting to death is seen within Jainism.

ARTICLE 2

THE MORALITY OF SALLEKHANA: THE JAINA PRACTICE OF FASTING TO DEATH

By Kim Skoog

The Jaina tradition is noted for many distinctive practices, though its adoption of *sallekhanā* or fasting to death is perhaps its most often noted aside from its observance of *ahiṃsā* or nonviolence. While not as widely practiced as sometimes reported, nevertheless, its practice has caused some discussion regarding its ultimate moral standing as the Jaina community has attempted to explain its moral and spiritual significance to non-Jainas. This paper will discuss the morality of *sallekhanā* by examining the practice itself and attempting to offer a conceptual schema to classify different type of suicidal acts. I will provide an ethical analysis of each of these types of suicides (including that containing *sallekhanā*) and see which types of acts are susceptible to traditional arguments for the immorality of suicide.

The word "suicide" is often used in a highly inflammatory fashion, often evoking negative connotations—containing religious and cultural biases. Jainas, for the most part, attempt to avoid calling *sallekhanā* suicide, and are defensive about making such identification as they themselves disapprove of nonreligious suicide; further, such an association they fear, would lead to condemnation of the practice on moral or religious grounds by outsiders. This tact is understandable, though unnecessary once one clarifies which class of actions is denoted by the term, "suicide." Clearly there are some praiseworthy and socially accepted acts that are classified as "suicidal;" further, with the increased influence of technology extending the length of biological life swith positive *and* negative results), many ethicists, theologians, and policy makers are rethinking the existing moral, religious, and legal views regarding "hastened" or "caused" death, e.g., doctor-assisted suicide and euthanasia.

Generally the term "suicide" is used in a lose and imprecise way, ranging from overtly selfish acts of self-destruction (estranged lover commits suicide to cause girlfriend guilt), to acts of self-sacrifice (suicidal act of hero), to self-destructive lifestyles ("drinking oneself to death"). This paper will continue to use the word "suicide" in this very broad fashion, though we need to attempt to articulate some sort of formal definition of suicide as the first step in devising a more sophisticated taxonomy of suicidal acts.

As a starting place, we can say that suicide is "the voluntary act of taking one's own life." Searching for its etymological origins, suicide comes from two Latin sources, the prefix, *suī*, or "of oneself," the suffix, *cīda*, "killer," which when put together means, "killer of oneself." So the suicide must involve—in its most essential meaning—a person who kills him or herself. The act must be voluntary in that if someone kills a person or forces the person to take his or her own life then the event is not suicide, but murder or execution. Further, defining the act to be voluntary rules out cases where the self-cased dead occurred accidentally—which again would not be a suicidal act. Still, let us dig a little deeper into the meaning of suicide.

Definitions of suicide frequently incorporate the ideas of *intentionality* and *self-imposed* death. The idea that a person committing suicide must do it "intentionally" carries much of the same requirement as our use of the term, "voluntarily." To be sure *most* people who commit suicide are aware that the act they are doing or are about to do will cause their own death, and *intend* to bring about that end by doing the action. Yet, some suicidal acts, ostensibly, do not require the rather demanding requirement of intentionality toward one's own death. For example, a person who is engaged in an act of self-sacrifice, e.g., standing in front of a person to "take" the bullet that is being fired at this other person, knows that by doing this action, she is placing herself at grave risk; yet, she is not doing it with the *intention* of dying; in fact, she may hope that the bullet will not strike a vital organ or that the bullet might miss her (and the intended target) altogether. The self-sacrificing person is acting voluntarily and is doing an act that likely will bring about her own death, but she is not intending to die, but only recognizes that by doing the act her death may result.[1]

Secondly, it is frequently stated in definitions and discussions of suicide that it is *self-inflicted* by the person who takes her/his own life. Again, as with intentionality, the vast majority of cases of suicide and attempted suicide involve someone inflicting him or herself with some instrument of harm that directly brings about his or her death. One causes oneself to die: it is due to his/her direct (inflicted) act that s/he dies. Yet, do *all* cases of suicide involve self-inflicted death? It is my position that the answer to this question is in the negative. We need to distinguish acts of commission from acts of omission. Clearly, the self-inflicted cases of suicide are of the former type, but don't we also have acts of suicide based on acts of omission? Take the case of a hunger striker in Belfast, Ireland who stops taking food as a form of protest against the policies of the British government. His act does not involve any direct self-inflicted acts, but rather the omission of the consumption of food. Yet it is clear that by doing this act the protester is putting him/herself at danger of dying of starvation and if he continues fasting, he will in fact die. The protester is aware of this fact, as are the authorities; otherwise he would not be committing this act of omission.

When we turn to a study of the Jaina practice of *sallekhanā*, we will see that both these terms (intention, self-inflict) will come to the forefront, as many Jainas argue that those engaged in this sacred practice are neither "directly causing (inflicting) their death" nor are they "actually intentionally trying to die." Yet since in our analysis of the meaning of the word "suicide," neither of these factors is considered a necessary condition for a suicidal act, *sallekhanā* could still be considered a suicide even if it was neither intentional nor self-inflicted.

Before turning to our main discussion of *sallekhanā*, let us briefly layout a taxonomy that subdivides these suicidal acts into different categories. My efforts to categorize different "types" or "categories" of suicide are based on either psychological or contextual factors present before or when the act is committed. My classification of suicide will be composed of three categories: Category I, egoistic suicide; Category II, altruistic suicide; and Category III, transcendental suicide.

Category I: Egoistic Suicide (Personal)

Egoistic suicide occurs when one chooses to takes one's life as a means for self-benefit. The principal defining characteristic of Category I suicide centers on motivation, one is engaged in a self-destructive act to benefit oneself primarily. Within this category, one might further distinguish between the kinds of motivation employed, malicious and non-malicious—the first immoral while the latter is not.

Malicious motivations, one's that inflict harm on another spitefully, may be viewed as immoral: killing oneself is done principally to cause harm to others, i.e., create guilt, great experience of loss, embarrassment, etc. In contrast, non-malicious, egoistic suicide is more likely to be view as morally acceptable, where one takes one's life solely to avoid uncontrollable pain, misery, indignity, etc.

Current views of society tend to favor an individual's right to choose to bring about their own death given that the pain and debilitation is intolerable with no prospect of relief—even if there is medical technology to maintain life further. It is interesting to note that the Jaina tradition does not support this basis for suicide. Undergoing continual pain is no grounds for suicide except within the context where the pain and disability would prevent observance of one's vows.

Category II: Altruistic Suicide (Nonpersonal)

Altruistic suicide occurs when one gives up one's own life to benefit or save others who are at risk of death or severe injury—one takes one's life free from personal self-gain. There have been countless cases of individuals who have heroically given their lives for the benefit of others. Consider a soldier falling on a hand grenade to save his colleagues, a Buddhist monk who engages in self-immolation—igniting himself as a protest against an unjust and oppressive government, or a hunger striker who fights for freedom and individual rights of a people. Presumably these are individuals who are acting in a self-less fashion (there is no or minimal individual benefit) and who are benefiting other people.

Altruistic suicide is defined by its relatively non-individualistic basis. Sometimes this situation arises spontaneously (falling on the grenade, "taking" a bullet for another, etc.); sometimes altruistic suicide is an act that is premeditated (Buddhist monk's self-immolation or hunger striker in Ireland). Yet in either context, one does one's actions principally for the benefit of others, one is sacrificing for the good of others with minimal or no benefit for one's self.

Without exception, all traditions (western or nonwestern) see this sort of suicidal behavior as moral, and in fact commendable. While both category I and category II types of suicide are viewed as voluntary acts leading to death and most are justifiable (*sans* malicious variety), nonpersonal, altruistic suicide is generally regarded as a superior, morally praiseworthy act.

Category III: Transcendental Suicide (Transpersonal)

This category of suicide is unique and apart from the mundane forms of suicide (I & II) in that it specifically relates to a transcendent or spiritual agenda: a distinctive perspective on life different from ordinary mundane existence. It conflates the personal/nonpersonal distinction, as both domains are satisfied albeit transcended (transpersonal). In category III suicides, the person's focus or orientation is "shifted" to a different level of existence or living, whereby one's own personal life concerns are transcended, and one is now solely identified with a supramundane agenda, i.e., spiritual liberation, ritual, calming of the mind, omniscience, etc. The Jaina practice of *sallekhanā* is an example of Category III suicide. This act can serve to complete the spiritual development of the individual, while at the same time allow for the "transcendence" from all egoistic, individualistic concerns. Other examples of Category III suicide would include *hara-kiri*, the Japanese practice of ceremonious suicide in serving the Bushido code of honor, and religious martyrdom, where an individual dies in order to preserve and/or promote a religious agenda. The following diagram offers a summary of these three categories of suicide and the relationship between them:

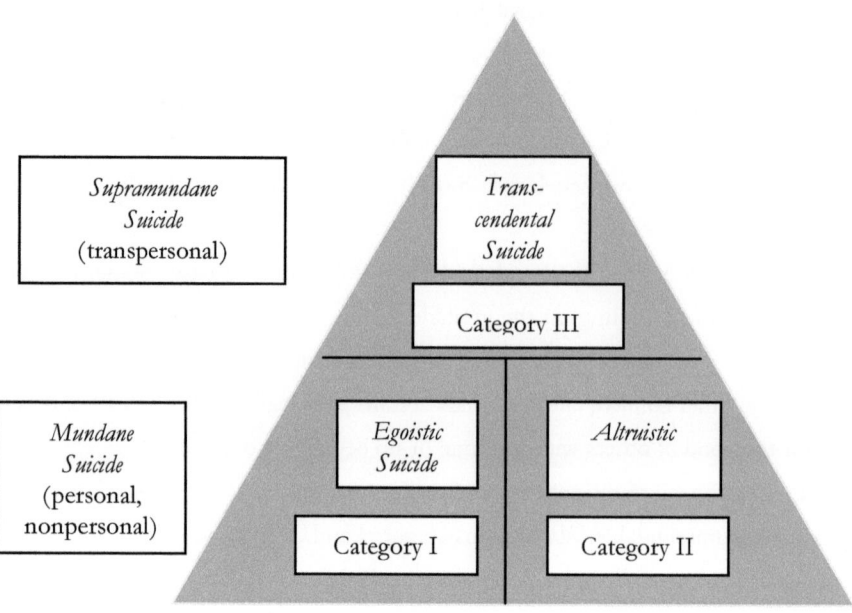

Let us now look at *sallekhanā* in some detail as a first step in trying to consider its moral status.

The Jaina Practice of *Sallekhanā*

Sallekhanā or Holy Death can be seen as an integral part of the Jaina Path. Stated baldly, it is the gradual fasting till death of an aspirant seeking liberation (*mokṣa*) within the Jaina tradition. The practice originally was done principally by mendicants at their own discretion. However, through the evolutionary development of the practice, restrictions were put in place, no doubt in part to help regulate its expanded usage by laypersons as well. Within the tradition four reasons or circumstances are mentioned that qualify for permission to begin the *sallekhanā* process. First, someone who has contracted a terminal illness and whose death is imminent (*niḥpratīkārā rujā*) may perform *sallekhanā*; second, a person who will likely be unable to perform one's vows due to dwindling capacities from advanced years (*jarā*), i.e., immobility, poor eyesight, forgetfulness, mental incompetence, etc. can start to fast till death; third, when there is a great shortage of acceptable food (*durbhikṣā*), such as during a famine, a person can stop eating; and fourth, when a person is thrust into a terrible calamity (*upasarga*) where one will compromise one's vows (e.g., severely wounded in battle or captured by the enemy), fasting to death is also permissible.[2]

The Jainas distinguish between pure or acceptable grounds for initiating a process that will eventually end one's life, practiced by a wise man (*paṇḍiya*) and that act or process that ends of life that is impure or unacceptable and done by a fool (*bāla*). In regards to the latter expression of suicide, Mahāvīra is unwavering in his condemnation of suicide that is brought about by drowning, burning, poisoning, use of a weapon, jumping off a cliff.[3] Presumably one reason for this condemnation is the inherent violence in these acts, not only to oneself, but also to the other living beings that might be harmed by one's violent act. Additionally, these latter types of acts involve extreme passion that again "contaminates" the purity and spiritual merits of the action. In contrast *sallekhanā* does not harm another being; in fact, it can and should be considered a supreme act of nonviolence, as one is no longer harming another being through the process of harvesting, storing, cooking, and consuming food. Further, *sallekhanā* requires dispassion on the part of the person who is allowed to engage in the process. Part of the screening process by the religious supervisor involves deciding whether the practitioner is truly ready to take on the long process of fasting to death and has had enough experience with fasting through his/her life to deal with the pain involved. Usually the process first involves cutting down on food gradually until one is eventually only consuming water every day. It is not meant to be a radical or swift exit, but rather a ritualistic and solemn spiritual practice. Depending on the aspirant's physical and mental condition, it can take weeks, even months to perform before death occurs.

In terms of what the Jainas would call "spiritual advancement," there is a wide range of individuals who partake in the practice. Some laypersons may take the vow of *sallekhanā* (*sallekhanāvrata*) long before they would actually commit to the practice, in the chance that if they might become ill or near death suddenly, then at least they can fast their last days, retake the five sacred vows, confess to their harmful deed and repent, and finally die in a meditative state.[4] But others, particularly mendicants or monks who have advance along in their spiritual quest, may approach and experience the process in a much more detached and embracing fashion.

The Jaina texts articulate fourteen steps or stages of spiritual purification (*guṇasthānas*) that are required to attain liberation (*mokṣa*), freedom from transmigration, suffering, and ignorance—realities inherent in all life. Key to this development is the experience of first having the correct view (*samyakdarśana*). Here the soul (*jīva*) first glimpses it true nature and shifts it perception on existence, no longer being preoccupied with its possessions, body, and worldly-orientated psychological states. The soul thus becomes detached from passions and attains a peaceful state (*viśuddhi*). Clarity is attain such that one now knows the forces that have bound and confused one for eons of lifetimes as well as the true nature of oneself. Hence, begins the long but rewarding process of attaining liberation. This orientation is particular important to recognize for our discussion in this paper, as it testifies to the alleged psychological orientation the aspirant has when engaged in the *sallekhanā* process. It can be observed that though the mendicant or layperson will benefit immensely from the practice, he or she personally is detached from the whole process: there is no deliberate motivation or intention toward some selfish or even egoistic goal. Further, even the desire for the goal of either liberation or the use of *sallekhanā* as a means to that end, have evaporated from the practitioner's mind. She or he is simply continuing to do a spiritual practice (fasting) that has been a part of most of his/her life; the difference in the case of *sallekhanā*, however, is that s/he is now taken a stricter vow that requires the eventual stoppage of all intake of food.

The practice of *sallekhanā* yields many benefits. The aspirant benefits (though not egoistically) as s/he progresses toward or attains liberation. Humanity, the animal and plant world, and the environment itself benefits from the influence of a being that has reached or become closer to enlightenment—through the teaching offered and the strict observance of non-violence. The process of *sallekhanā* is rooted in the transcendent basis of the process leading to liberation and the Jaina focus on contemplation and self-growth.

In summation, *sallekhanā* is a spiritual practice directly leading to death but stands apart from either egoistic suicide or altruistic suicide. It is distinct from the former (egoistic) in its lack of selfish or beneficial motivation behind the act, while it differs from the latter (altruistic) in its lack of a purely selfless orientation. Transcendental suicide "goes beyond" the day-to-day, mundane mode of action, getting its motivation instead from a spiritual orientation. Having stated some basic information about this practice, let's consider reasons (arguments) for the immorality of suicide and see if and how they might apply to the aforementioned categories of suicide.

Arguments Against Acts of Suicide (Western & Asian)

Judo-Christian-Islamic position

The Near-eastern traditions all adhere to the view that God's provident plays a fundamental role in the unfolding of history, thus any intentional deviation from that plan is devious, immoral, and sinful. Islam is clear in its condemnation of suicide, "Do not kill yourself as God is merciful to you." (*Koran* IV: 29, 30) Such acts bring disgrace to one's family and result in hell for the person doing the act. Christian and Jewish traditions generally have been against suicide, though ambiguity on this issue is present in the first few centuries. However, from Augustine onward, the message against suicide is undeniable, as the following arguments make clear.

Theological Argument

Life is a gift from God, each life has been created for a purpose according to God's plan. If we elect to kill ourselves before our life has played itself out—with all the events that God has envisioned for us—then we violate God's will, sin against God, violate the natural order. We "desert our post," not upholding our role in the divine plan that God has setup—(phrase used by Socrates, Aquinas, Kant.) Thus, suicide is immoral.

Natural Law Argument

All living beings have natural inclinations toward self-preservation—everyone loves oneself. Suicide, *de facto*, is self-inflicted destruction toward oneself, hence contrary to natural instincts. Proper behavior, acting morally, is co-extensive with natural inclinations. Thus, suicide is immoral.

Social Argument

Suicide naturally does harm to other persons, society at large. For killing oneself does injury to the community, it precludes further contributions to society by that individual. Thus, suicide is immoral.

Asian Positions and Suicide

In spite of the fact that Jainas practice *sallekhanā*, there is condemnation of (egoistic) suicide by Jainas and within Asian traditions in general.

Hindu–Buddhist–Jaina Views

A basic cause for the prohibition against suicide stems from their presupposition of transmigration and liberation. Suicide removes the opportunity to burn off past stored karma and conduct practices that can lead to liberation; for while suicide might remove one from immediate pain and suffering, it is based on a short-sided perspective, since one will be reborn again into the world of suffering.

Liberation Argument

Human life is very precious: rebirth as a human has the most potential for moving toward liberation. If one kills oneself, one ruins this valuable opportunity. Hence, suicide is counter to perceived spiritual goals. This argument is more of a pragmatic observation than a theological or eschatological scolding. Note that there are some "threatening" passage in the literature, e.g., "He who takes his self (life) reaches after death, sunless region covered with darkness." (*Īśāvāsya Upaniṣad*)⁵

Non-violence/Compassion Argument

Jaina and Buddhist teachings emphasize nonviolence and kindness, so killing oneself is exhibiting a lack of these two traits to oneself. It is wrong to harm another living being, and one incurs demeritous karma as a result of such acts. Accumulation of demeritous karma serves to retard spiritual growth. Suicide does harm to a living being, i.e., oneself. Hence, Suicide is immoral and counter to spiritual advancement.

Response to these Arguments from the Jaina Perspective

These arguments were initially raised against what I have termed egoistic suicide (category I). They collectively argue that selfish, self-serving acts of self-demise are immoral because of the sorted harm they inflict. But, do they apply as well to category III suicide? Consider *sallekhanā* as a paradigm case and its susceptibility to the arguments given above.

Theological Argument

One can observe first that Jainas do not recognize the providence of a monotheistic God, so would not be concerned about "violating" a divine plan. Secondly, within the Indian context, as gods recognize the principal importance of liberation, any act that can dramatically speed up the attainment of liberation (such as *sallekhanā* can do), would surely not be against divine wishes or goals.

Natural Law

From the Jaina point of view, the perceived "normal" human condition is itself confused or ignorant, and needs correction and alteration. Our so-called "instinct" the natural law argument identifies, i.e., "clinging to life," is actually an impediment to personal spiritual development and must be controlled and eventually quieted for spiritual progress to occur. New, spiritually inclined "instincts," i.e., seek liberation, avoid harm to others, should be incorporated into life, and *sallekhanā* is totally in conformity with these goals.

Social Responsibility Argument

Jainas delineate a hierarchy of social values and obligations or vows (*vrata*). The highest value is to attain liberation, so there is no better act to oneself or society than to strive toward liberation.

If one decides to renounce the world and through one's asceticism attains to a higher level of spiritual development that helps thousands or even millions of beings to move closer to liberation, then surely we can give no greater contribution in our social role.

Liberation Argument

This argument is a non-starter, for if this particular act (*sallekhanā*) brings about liberation (along with death), then it is not in violation of the goals that form the basis of this argument against personal suicide. It is significant to note that this form of suicide does not directly benefit the individual in the same way that category I suicide does, for in category III suicides, the person "gains," but transcends any direct personal benefit—as usually defined. There is no selfish dimension to the act, and in fact, the individual ceases to be the self-centered egoist that he or she once was. Category III suicides lack the selfish or personal (beneficial) motivation behind the act that causes the immoral categorization of some category I suicides. Yet, there is still a sense of "individual" advancement or compliance that distinguishes this category from category II suicides.

Ultimately, the moral status of *sallekhanā* can be determined by whether it violates any prevailing moral prohibitions based, presumably, on one or more of the arguments covered above. These arguments center on concerns over spiritual, social, and personal transgressions. As was just argued, *sallekhanā* does not violate any of theses areas; but to the contrary, *sallekhanā* consummates the ideals of these notions. *Sallekhanā* is the ultimate act of self-expression, giving rise to great spiritual, (trans) personal, and social advancement and improvement.

Notes

1. "Intentionally" is a commonly included attribute of suicide. Many suicidologist would not agree with my decision to disallow its inclusion as a necessary condition of suicide; still others would not include fatal acts of self-sacrifice as cases of suicide. Joseph Margolis (1975), for example, excludes cases of self-sacrifice as being suicide because the intentions of such a person are not directed toward ending his or her life; Tom Beauchamp (1993), in contrast, does allow for certain self-sacrificing people who bring about their own death to be included as suicide, so long as they intend to kill themselves.
2. These four permissible situations for the initiation of *sallekhanā* are catalogued in the *Ratnakaraṇḍaśrāvakācāra* of Samantabhadra as mentioned in Jaini 1979: 229.
3. Schubring 1962: 289.

4. *Tattvārtha Sūtra* 7.17 state, "The householder should become a practitioner of the penitential rite of emancipation of the passions bye a course of fasting which spans a number of years and ends in death." In the Śvetāmbara commentary of allegedly Umāsvāti himself entitled, *Svopajña Bhāṣya*, on the passage, Tatia (1994: 178) paraphases it, "The practitioner starts by reducing his diet, then fast regularly for progressively longer periods, adopts the observance of the ascetics's self restraint and finally gives up all food and drink to fast to death while engaging in reflection ... and meditations. ..." In the Digambara's accepted commentary by Pūjyapāda Devanandi called the *Sarvārthasiddhi*, Tatia (1994: 178–179) again paraphrases, "The rite of emaciation is undertaken by the household for the attenuation of the external body and passions. It is adopted with full joy and calmness of mind and not impetuously. It is not suicide because it is undertaken without duress or passion. To commit suicide is to kill oneself out of anger, agony, malice or frustration, whereas fasting to death purges the should of its passions and perversities by conquering the fear of death."

5. As quoted by Rao 1983: 211.

DISCUSSION QUESTIONS

1. The ritual of fasting to death is seen as an observance to what concept?
2. What two forms of criteria are used to decide whether a death is considered a suicide?
3. Summarize and then compare and contrast these types of suicide: egoistic, altruistic, and transcendental.
4. Explain the practice of *sallekhana* in Jainism. List three points you learned from the reading.
5. Compare the arguments against suicide from the various traditions mentioned in the article.
6. What are some Jain responses to these arguments, and do you feel they are justified? How should these responses impact the future legality of this ritual within Jainism?

WOMEN

There is a lot of feminist criticism about the large divide between the characteristics of the goddesses within the Indian traditions compared to the characteristics of ordinary women. According to Whitney Kelting, goddesses in Jain literature typically present positive images of the feminine. Kelting notes that although this is true, none of the Jain women she spoke to mentions this connection with the goddess as a part of their understanding their own sense of empowerment. This article looks at the understanding of the feminine in Jain literature, and it allows us to ponder what that could mean for Jain women.

ARTICLE 3

CONSTRUCTION OF FEMALENESS IN JAIN DEVOTIONAL LITERATURE

By M. Whitney Kelting

Fasting and Female Narratives

The virtuous women (*Mahāsatī*)[1] serve both to exemplify an alternative to being a housewife and to give laywomen a way of situating their personal renunciations (usually fasting) in a narrative structure. There are several Jain fasts that a woman might do to fortify her position as a married woman and to protect her children (*saubhāgya*). The perfection of the soul which can be gained from austerities leads the Jain woman to powers which can control both domestic and spiritual matters. Fasting not only gains a woman merit, protects her family from harm or illness, helps a woman get a good husband, assists her family in financial success, gains prestige for her family or proves a woman (and her daughter's) honor, but it is also an important way for women to attain a piece of the perfection usually reserved for nuns or Satīs. After a woman has completed a major fast, she is reverently displayed in much the same way a religious image—for Jains, the Jina image—is: in parades accompanied by bands and dancers, in celebrations with offerings made to her. This way, pious women also become a public icon of Jainism expressed to Jains and non-Jains.[2]

The central narrative from the *Śrīpāl Rājā no Rāsa* and the *Navpad Oḷī* fasting texts conjoins the worldly and spiritual merits of fasting: A great king, Prajāpāl, was hostile to his daughter, Mayaṇāsundarī's devotion to the Jinas. The princess, Mayaṇāsundarī, felt that her good fortune came from her devotion rather than from her father's power and she said so in public. The king was furious and in order to assert his authority over her, he married Mayaṇāsundarī to the leper, Śrīpāl. She continued her daily Jain worship and prayed to Cakreśvarī, the guardian goddess of Ādināth, to heal her husband. Cakreśvarī told her that all would be well soon, and the next day Mayaṇāsundarī met a Jain monk who suggested that she perform the *Navpad Oḷī* fast to cure her husband. Mayaṇāsundarī did perform the fast and on the ninth and final day of the fast Śrīpāl and 700 other lepers who were in the area were cured. Later Śrīpāl, having learned this fast from his wife, gained great wealth as a result of his devotions.

Mayaṇāsundarī's fate and fortune, though temporarily set back by her father's rash actions, were not significantly deterred, because of the power of her devotion and asceticism.[3] Even Cakreśvarī does not simply heal Śrīpāl; the goddess reassures the princess and perhaps even sends the monk to the princess. It is the princess who, through her religious discipline, determines her fortune and heals her husband. Likewise the wife here shows her husband how to perform the fast and teaches him the worship which leads to their family's wealth and prosperity. Cort writes that during the *Navpad Oḷī*, Jain monks gave sermons telling the story of Śrīpāl's worship rather than Mayaṇāsundarī's.[4] While laywomen always cited Mayaṇāsundarī's success when they talked about the fast's efficacy suggesting that laywomen performing this fast are following Mayaṇāsundarī's devotion in hopes of protecting and uplifting their families. Though this fast has positive effects on one's karma, one can see how the merit accrued is channeled into worldly gain. The moral, if there is one, is that a woman can defy her father's wishes and curses and overcome adversity by being a pious Jain and by following the advice of Jain monks.

In the other most commonly retold fasting narrative, the heroine, Candanbālā, has not chosen to perform most of her meritorious acts; it is her choice to put her piety before the resolution of her troubles that ultimately earns her *mokṣa*. Candanbālā was a beautiful princess who was jailed and then released into a merchant's family as a maidservant. The merchant's wife was jealous of her beauty and when the merchant went away on business, she shaved Candanbālā's hair off (hair being a common symbol in India of a woman's sexuality) and bound her in chains. The merchant's wife left her for three days without food and the merchant returned. He was horrified and released her. There were only black lentils (*uḍad ḍal*) in the kitchen, which he offered her, but she wouldn't eat until she had offered alms to a Jain monk. Candanbālā sat at the house's threshold reciting the *Navkār Mantra* and sorting lentils. Then Mahāvīr passed by the house. He had been fasting for five months and twenty-five days because no one had met his conditions for fast-breaking: That an unmarried girl from a royal house who now was a servant, having had her head shaved and fasting for three days would be sitting in a threshold sorting lentils saying the *Navkār Mantra* (the basic statement of faith for Jains) and weeping. Candanbālā fulfilled all these conditions but was not weeping. When he passed by and wouldn't accept her alms, she began to cry. Mahāvīr came back and took alms from her. Candanbālā's hair immediately grew back and she forgave the merchant's wife who had by her cruelty allowed Candanbālā to give the fast-breaking alms to Mahāvīr. Later Candanbālā became a nun ordained by Mahāvīr and attained *mokṣa*.[5]

The final verse of a popular devotional hymn (*stavan*) tells an abbreviated version of the story of Candanbālā:

Candanbālā was a princess who did the three day fast.
Bring black lentils in the winnowing fan, that alms-giver, Mahāvīr, and that trader's wife attained *mokṣa*.

The hymn's abbreviated version of the Candanbālā story reminds the hearer of the full details of the narrative in a succinct form. It suggests that torments may offer an opportunity for great merit. In fact the hymn, through its earlier call for practicing extreme fasting, suggests self-imposed austerities modeled on Candanbālā's imposed fast. This three-day total fast is called the *Aṭhṭham Tap* in the hymn. The Aṭhṭham Tap is a famous fast said to have been performed first at Śaykheśvar (a central pilgrimage site) by Kṛṣṇa in order to free his army from a curse.[6] Now this fast is performed for the wellbeing of one's family and especially one's husband (*saubhāgya*), not for expiation, by many Jains—often in imitation of Candanbālā's three day fast which resulted in her worldly result—symbolized by her hair growing back and the acceptance of her faith by the town, and spiritual result—she later attains *mokṣa*. Candanbālā attains *mokṣa* because of her self-imposed austerities which result in the intimate connection—that of being worthy of giving the Jina his fast-breaking meal—with Mahāvīr. In Mayaṇāsundarī's case, it is defending her own faith that gains her *mokṣa*; while in the Jinamātā case it is the fulfillment of expected motherly duties which leads to their *mokṣa*. In both cases, these stories serve contemporary Jains as allegorical models for understanding women's lives and goals.

Satīs and Faithful Wives

All the women I knew were familiar with the rich and extensive initiation (*dīkṣā*) narrative tradition in which the Jain layperson—almost always female—renounces the world. Additionally, most families had at least one relative in living memory who had become a nun. It is important to note that as opposed to Hindu traditions, in which female renunciation is neither socially acceptable or necessarily a viable option,[7] Jain women can choose to be nuns. This choice seemed far more salient than narratives of Mallināth, the theoretical arguments about women's equality in the spiritual realm or what can be garnered from goddess devotion. The reasons for becoming a nun are usually a combination of a variety of socio-economic conditions, the personal ambitions of the woman, and her attitudes towards and prospects of marriage.[8] Mahāsatī narratives provide compelling narrative incentives to adopt the prestige of the nun's life—or those practices associated with that path—over the social realities of an ordinary woman's life.

Among the numbers of Jains worthy of note and honor in Jain rhetoric, the sixteen Satīs (virtuous women) stand out as especially ubiquitous. These Satīs are often listed as follows: Brahmī, Sundarī, Candanbālā, Rājīmatī, Draupadī, Kausalyā, Mṛgāvatī, Sulasā, Sītā, Damayantī, Śivādevī, Kuntī, Subhadrā, Śilavatī, Prabhāvatī, and Padmāvatī. These women are honored because they stayed true to their religion when they had the most to lose by doing so.[9] While the Satīs are not equally well known, nor do they have equally detailed narratives, several of them have found their way into the fasting and other devotional vernacular literature. The Satīs whose stories I heard most often were Candanbālā, Rājīmatī and Subhadrā, but one cannot help but notice the apparent appropriation of Hindu goddesses—Sītā, Draupadī, etc.—into the list. These Hindu queens usually end up taking initiation as a result of having seen a Jina or having heard the Jina's lectures. These spontaneous conversions are a common part of the pan-Indian strategy of proving the spiritual superiority of one path over another, well-documented in the study of Hindu literature.[10] The narratives of Sītā and Draupadī, the heroines of the *Rāmāyaṇa* and the *Mahābhārata*, are "converted" and then deemed worshipful in their position as Jain Satīs.

The Sītā story has some of the standard Hindu version details about her family and lineage but quickly takes a Jain turn. As opposed to the Hindu (both the Valmiki and Tulsidas) versions, the point of the satī narrative is not to prove Rām's perfection but Sītā's. The conversion of Sītā shows how the Jain ideals of womanhood are applied to the epic heroine; it highlights the central features of Jain womanhood. The version in the short tract on virtuous women, *Sulāsā Candanbālā*, is based on the longer Jain Rāmāyaṇa, Vimalasūri's *Paumacariya*.[11] In her youth Sītā was impressed by the Jinas, Jain monks, the beauty of the Jain religion, and the truth of the Three Jewels while her friend did not show any respect to either the Jina or the Jain monk. (This allows for Sītā's later conversion by her having already had the "Correct Faith.")[12] Then Sītā is wedded to Ram. The whole story is cleansed of Rām's violence, including the famous story of Sītā sending Rām to capture the golden deer leading to her capture by Rāvaṇ. The story does include a variety of fire tests for Sītā to prove her loyalty to her husband Rām. Finally, a Jain monk arrives and tells Sītā the story of a holy woman who took initiation, performed extreme fasts, went to heaven and was born as Sītā. Sītā then understands the point of the monk's story and converts to Jainism, starts living a Jain life and goes, when she dies, to live in the abode of the gods. Rām then becomes a Jain monk. Later, as a result of her virtuous acts, she attains mokṣa.

Here Sītā, without in any way violating her image as the ultimate faithful wife (*pativratā*) attains *mokṣa*, because in a past life she was a Jain nun. For Sītā to have done so in this life, she would have had to either outlive her husband (then as a widow, she would have failed in the ideal of the perfect wife) or renounce her husband (but a true "faithful wife" always puts her husband before her and thus would not pursue her own spirituality if it meant he had no wife to care for him). While most studies of *pativratā* rhetoric focus on Hindu women,[13] socially, Jains also expect a woman to take care of her husband, eat after he is done, subordinate her wishes to his and to die first. However, and this is extremely important, the ideal of Jain womanhood is to be a nun—a woman who renounces family life in exchange for a spiritual one. If she so chooses, and she is not under eight years old, pregnant or the mother of small children, and her parents or husband agree, a Jain woman may become a nun. Though I have not yet heard of a woman who was actually prevented by her family from becoming a nun, the rhetoric of initiation narratives often includes the resistance of her parents and family; the woman's steadfast dedication ultimately succeeds in her initiation. Most of the Satīs take initiation as a final act in the narratives of their pious lives. The Satīs all achieve *mokṣa*, sometimes after a lifetime in the abode of the gods. However, their lives as nuns are rarely elaborated upon. The piety of the Satī is tested and proven as a laywoman. The test complete, she becomes a nun or goes to the abode of the gods or attains *mokṣa* right there.

The best known exception to this pattern is in the story of Rājīmatī whose narrative extends beyond her initiation. The narratives of Rājīmatī I heard women tell were often short pieces deriving from the longer Sanskrit and Old Gujarati narratives about the life of the 22nd Jina, Nemināth.[14] Rājīmatī was engaged to Nemināth and he renounced the world. Rājīmatī's brief life as a laywoman was pure; when her friends lamented that Nemināth was taking initiation, instead of joining them Rājīmatī decides she too will take initiation. Rather than realigning her loyalty away from her fiancé she, too, renounced the world and became a nun. Rājīmatī not only proved her loyalty and her morality through this story but later, too, when Nemināth's brother tried to seduce her she gave him a lecture in which she describes herself as the vomit of Nemināth because she sees herself as effectively married to Nemināth. Nemi's brother then became a monk. Here, Rājīmatī, like Sītā, followed a *pativratā* ideal even into initiation. After initiation, chastity and modesty are associated with one's own soul and no longer with one's relationship to the now renounced husband. Additionally, Rājīmatī attains *mokṣa* before Nemināth, thereby like an auspicious wife that she never got to be, she dies before her would-have-been-husband. Because her fiancé renounces before her, Rājīmatī's entire career is untouched by inauspiciousness or taint. As opposed to Sītā—whose Hindu husband cannot be urged even to be Jain –, Rājīmatī is permitted by her husband's renunciation to renounce and ultimately attain *mokṣa* in this lifetime. Thus the narrative stresses that to be a nun is the most direct path to *mokṣa* and a symbol of the best a woman can achieve in this lifetime; however, they suggest that the choice of being a nun is available only after fulfilling one's duties as a wife.

However, the popular narrative of the Mahāsatī Subhadrā challenges the model of the ideal wife. As told in the tract, *Sulasā Candanbālā,* Subhadrā refuses to convert to her husband's religion, Buddhism, an act which goes against the expected norm for a new wife. Instead Subhadrā continues to go to the Jain temple against her husband's wishes. When her mother-in-law orders her to go to the Buddhist temple, Subhadrā locks herself in the house. Later Subhadrā, seen removing a splinter from a Jain monk's eye with her tongue, is assumed to be unchaste. She prays to the guardian goddess who bolts the city gates. The townspeople ask how they can open the gates. The goddess says that when a true woman (*satī strī*) draws water in a sieve tied with a weak thread and bathes the doors, they will open. Everyone tried and failed. When Subhadrā says she can do it, her mother-in-law refuses saying that she is no *satī*. Subhadrā says that she will have the goddess open the doors. The goddess opens three of the four doors and says another *satī* will open the fourth which Subhadrā does. After this miracle, everyone takes Jain initiation and attains *mokṣa*. It is Subhadrā's devotion that garners the attention of the goddess and ultimately converts and leads to the *mokṣa* of all the townspeople.

This story is a good narrative to illustrate for women the importance of keeping faithful, especially if they are married into a family with less religious practice or a different religious practice than that with which the woman grew up. The message is that through steadfast piety a woman can lead to the betterment of her husband's family and town. Thinking back to Sītā and Rājīmatī whose stories uphold the social expectations about women's behaviour we can see how the Mahāsatī literature does support social norms. However when those norms challenge the woman's Jain religiosity, the bottom-line is that being Jain is more important than being a good wife in any generic, pan-Indian sense, and that struggle against social norms makes the Mahāsatī all the more holy.

While this is a preliminary examination of these Jain devotional narratives, I think they do suggest that the understanding of women's perfection (or lack thereof) that scholars have found—in Hindu traditions and in the study of goddesses and feminine imagery—will have to be expanded and reoriented. The new model will have to confront contemporary Jain rhetoric about ideal women and *mokṣa* which places human women at the centre of religious perfection. In popular, contemporary Jain narratives—and in many of the older root sources—perfection, more often than not, takes the form of mothers, daughters, wives, and nuns.

Notes

1. Though this use of this term suggest the women's unending devotion to their husbands, I understand the term in its more generic form "a chaste and virtuous woman" or "one knowing the truth" rather than the specific Hindu understanding because not

all *satīs* have been married (notably Candanbālā). While Somani (1982: 79–80 in Dundas 1992) found evidence of Jain women performing the rite of self-immolation on their husband's death, contemporary Jains see this as suicide and therefore an irredeemable act of violence. The relationship between suicide and *satī* is as tricky for Hindus (Hawley 1994; Mani 1989; Weinberger-Thomas 1999) as suicide and ritual fast-unto-death issue is for Jains (Tukol 1976). This term may have been chosen to encourage Jains to worship virtuous Jain women rather than women considered virtuous by Hindus. I am presently researching this complex interaction.

2. Kelting 2001.
3. The story of a wife whose unswerving devotion saves her family is by no means unique to Jainism; there are many such stories within the Hindu tradition as well.
4. Cort 2001.
5. The Candanbālā story as retold here is based on both the version in a tract about virtuous women (*Sulasā Candanabālā*) and retellings I heard in the contexts of both the Candanbālā Fast (the three day fast whose fast-breaking reenacts the Candanbālā narrative) and interviews with Jain laywomen. They match with the versions told to Reynell which she records in her thesis (1985: 239–241).
6. Cort 2001.
7. Ojha 1981.
8. Holmstrom 1988; Reynell 1985 and 1991; Shāntā 1985.
9. Harishbhadra 1993. Other lists are similar but often replace Śilāvatī with Puṣpacūlā.
10. See for example, Harlan 1992; O'Flaherty 1988; Ramanujan 1973.
11. This text can be dated somewhere in the 1-5th centuries CE.
12. Most women I knew believed that once a person has the "Correct Faith" (a belief in the truth of Jainism) they will eventually attain *mokṣa*. This is their reformulation of the question of whether a soul is *bhavya* (capable of *mokṣa*) or *abhavya* (eternally incapable of *mokṣa*) which is used in Śvetāmbar-Digambar debates over whether or not Makkhali Gosāla was *bhavya* (Ś.) or *abhavya* (D.), (Dundas 1992: 90). The women's reformulation probably hinges on their understanding of *bhavya* as one who has some attraction to the Jain faith.
13. Harlan 1992; Leslie 1989.
14. See Vaudeville 1986—which includes a translation of the *Rājal-Bārahmāsā*—, the *Triṣaṣṭiśalākāpuruṣacaritra*, Vol. 5 and the *Uttarādhyayana Sūtra* 22.14–19.

DISCUSSION QUESTIONS

1. What types of fasting might a wife practice? Why would she perform this fasting?
2. What is the story about the Princess Mayanasundari? What lessons about fasting can be learned from her story?
3. Describe the role of Mahavira in the story.
4. How does Princess Mayanasundari attain *moksha*?
5. Can Jain women choose to be nuns?
6. Explain the Jain version of the *Ramayana*? Describe Sita's role in the story.

JUDAISM

INTRODUCTION

Abraham's tomb is located in the city of Hebron; today this location is within the occupied territory of the West Bank. Many of the sacred sites for the Jewish faith are located in Israel proper and in the West Bank. These sacred sites connect believers with their sacred history. This physical connection is powerful, as the struggles of the past are still alive for the Jewish people. In Judaism, Abraham is the father of **monotheism,** or "the worship and belief in one single G-d." (As an act of respect, religious Jews avoid saying Gd's name; this extends to all expressions except prayer and Torah study. Throughout this chapter, "G-d" will be used.) The sacred story details a covenant made between G-d and Abraham; it lists the promises G-d made to Abraham and everything that was to come for his descendants. One of these descendants was Abraham's grandson Jacob. One night Jacob wrestled with an angel—understood by many in Judaism to be G-d. Thenceforth, Jacob was known as Israel. This new name highlights how G-d ultimately prevailed throughout the struggle. Israel would have 12 sons, who became the leaders of the **12 tribes of Israel.**

In the Hebrew Bible, this community is known as the "children of Israel," and they still struggle with this G-d. Across the large spectrum of branches within one of the world's smallest religions, many Jewish people feel that G-d still prevails in their lives, especially in belief and practice. But how do they understand this G-d? In most forms of Judaism, the concept of G-d is not expressed anthropomorphically. **Anthropomorphism** is the concept of giving humanlike characteristics and traits to an entity that is not human—in this case, G-d. In most forms of Judaism, G-d is a spirit.

This spirit is without gender and is omnipotent, omnipresent, and omniscient. These concepts also reflect a transcendent form of G-d. **Transcendence** implies something "beyond"—in this case, G-d is outside of time and space. The children of Israel built two temples for their G-d, both of which were destroyed, the first by the Babylonians in 586 BCE and the second by the Romans in 70 CE. There has not been another temple since.

The temples were designed with specific sacred spatial boundaries. They were cared for by a priestly line connected to the tribe of Levi. (Levi was a high priest for the children of Israel.) These temples had specific areas and rooms, one of which was known as the Holies of Holies. On *Yom Kippur*, or the **Day of Atonement**, the high priest could access the presence of G-d in that room; this is a concept known as ***shekinah***, or "dwelling." This encounter, which was performed to seek G-d's forgiveness for the sins of the whole community, was achieved through the sacrifice of specific clean and unblemished animals. These sacrifices have not been done since the destruction of the second temple. Jewish people around the world feel the absence of their temple, but even without a temple sin must still be reckoned with. The Day of Atonement now includes self-sacrifice through fasting and prayer. Various rituals are performed, and people seek forgiveness from people whom they have wronged.

At the Temple Mount in Jerusalem—the only location specified for the building of the temple—stands the Dome of the Rock. (This is a sacred site for Muslims as well; for them it is the gateway to heaven.) All that remains of the second temple is the ***Khotel***, or "Western Wall." The wall is visited by large numbers of tourists, but most importantly it is visited by devout Jews and Christians, who pray at the site continually throughout the year. Some leave their prayers on small pieces of paper that they slip into the cracks between the stones. For many believers, G-d still intervenes in our lives, and prayer is a prominent form of communication with G-d. Judaism has many prayers, which must be offered with the right ***kavanah***, or "intention." In some forms of Judaism, the person who is praying wears specific religious garb. This can range from a ***tallit***, or "prayer shawl," to a ***tefillin*,** a small black leather box bound to the person's forehead and forearm (these boxes contain tiny scrolls with scripture written on them).

Scripture, sometimes referred to as the Hebrew Bible, has a special place in the heart and soul of the Jewish faith. In sacred story after sacred story, the struggles that the children of Israel had to endure in order to honor their G-d are elaborately detailed. Their struggles are countless, beginning with Abraham's arduous tests of faith. The Jewish people endured slavery, overcame idolatry, and fought through inescapable genocide. They wrestled against the seductive forces of power within the eventual monarchy and priesthood. Initially, the children of Israel were guided by a group of people collectively known as *Judges*. They instructed the community through prophecy and during times of warfare. This era came to an end when this community began a monarchy. Their second king, known as King David, stands out most notably from this time period. It was his son Solomon who built the first temple. The priesthood for that temple was then established. In the end, the scripture recounts the Jewish people's faith in G-d, even after living through destruction after persecution after destruction after persecution. The Hebrew Bible is also known as the *Tanakh*. In the Hebrew language, *tanakh* is an acronym (TaNaKh) for the three different parts of the scripture.

The first section of the *Tanakh* is the **Torah**, the first five books of the Hebrew Bible. In many branches of Judaism, these five books are believed to have been written by Moses himself. They contain the bulk of Jewish laws and commandments. There are 613 *mitzvahs,* or "commandments." The second section is the *Nevi'im,* or "Prophets." These books show the messages G-d brought to the children of Israel through messengers or Prophets. The Prophets were rarely liked, as they brought hard messages that aimed to shed a light on those who had dishonored and disobeyed G-d. Lastly, there is the *Ketuvim,* or "Writings," which include wisdom literature and poetic songs. These writings are immortalized at weddings and funerals across the Jewish and Christian faiths. Before the discovery of the Dead Sea Scrolls, the oldest manuscripts of these scriptures were dated to the medieval period.

So far, the **Dead Sea Scrolls** have been the most important finding for the study of the Hebrew Bible. The Scrolls contain almost all of the books of the Hebrew Bible, except for the book of Esther, and they had been preserved and untouched since the Second Temple Period. When the Dead Sea Scrolls are compared to the scriptures as presented in medieval manuscripts, they are virtually identical. The Hebrew Bible is studied daily by many in the ultra-Orthodox and Orthodox branches of Judaism. **Rabbis**—community teachers and leaders within the Jewish faith—also have a profound interest in the study of the scripture and of law. Rabbis help the community in practice and law, and after the destruction of the second temple they became essential for the preservation of Jewish tradition. They established a rabbinic academy in the same year as the destruction.

For many Jewish people, the Hebrew Bible is mysterious, full of inconsistencies and contradictions. Fortunately for people across all branches in Judaism, there exists the rabbinical literature known as the **Babylonian Talmud**. This Talmud is filled with *midrash*, or "commentary," on the Hebrew Bible. It is composed of two parts: the *Mishnah* and the *Gemarah*. The *Mishnah*, which means "repetition," was compiled by Rabbi Judah the Prince in the third century CE. Judah's intent was to repeat and preserve centuries of rabbinical discussion and commentary about the

Hebrew Bible. He divided the *Mishnah* into six sections, and it covers most aspects of law and practice. The **Gemarah** is further commentary on the *Mishnah*, and the page layout of the Talmud resembles a book within a book.

The whole of the Talmud teases out the most difficult questions about the Hebrew Bible, and it sheds light on the Bible's mysteries through discussion and analyses by many sages and rabbis. The medieval Jewish philosopher **Moshe ben Maimon**—also known as **Maimonides**, and sometimes referred to by the acronym **RaMBaM**—is one of the most famous Talmudic scholars. He synthesized all of the elaborate discussions from the Talmud and emphasized their conclusions. His 14-volume work, the ***Mishneh Torah***, is the universal and accessible reading of Talmudic law.

Many of the Jewish celebrations and holidays commemorate the sacred history of the Jewish people. The rituals connect believers to who they are, and to how G-d was—and continues to be—there for them. Perhaps one of the most powerful commemorations occurs during the festival of **Pesach,** or "Passover." The festival memorializes how G-d liberated the Jewish people from bondage in Egypt. It chronicles the story of Moses, and it shows the power of G-d in confronting the Pharaoh. This festival celebrates the freedom G-d brings, as well as everything that G-d provides. It is a story of total patience with G-d and complete dependence on G-d. Many Jewish people celebrate a Passover *seder*, a dinner in which all the prayers and food recount and represent specific details of the Exodus story. A great deal of symbolism and joy is expressed in remembering how the soul of the children of Israel thrived even through the most difficult of times because they relied on their G-d to see them through it all.

ECOLOGY

Zionism refers to "a political movement that sought to establish the historic and sacred land once known as Canaan—and known in more recent history as Palestine—as the rightful land and home of the Jewish people." Since the creation of the nation-state of Israel in 1948, the goal of Zionism is to protect this land. Through these efforts, the deeper concerns of environmentalism within Judaism became more evident. Although a variety of early Jewish thinkers, sages, and rabbis have written on the subject of ecology and Judaism, the topic has been revived in full force through many of the later tragic events like the Holocaust. The following article explores some of these connections.

ARTICLE 1

JUDAISM AND THE SCIENCE OF ECOLOGY

By Hava Tirosh-Samuelson

Contemporary Jewish Environmentalism

The Jewish religious tradition is replete with ecological wisdom, but Jews began to encounter this fact only in the 1970s, with the rise of a Jewish environmental movement (Tirosh-Samuelson 2006). It emerged as an apologetic response to the accusation of Lynn White Jr. that Judeo-Christian tradition was the root of the current ecological crisis. In defense of Judaism, Jewish theologians have argued that White and other environmentalists either misread the Hebrew biblical text, or lack any knowledge of post-biblical Judaism. An accurate and informed reading of Jewish sources shows that the Jewish religious tradition, and especially the Bible, can be the basis for sound environmental policies. Jewish environmental activists began to explore ways to integrate their environmental sensibilities with their desire to live a more meaningful Jewish life. Guided by Zalman Schachter-Shalomi's Jewish Renewal Movement (Schachter-Shalomi 1993), environmentally concerned Jews began to mine the Jewish literary tradition for its ecological wisdom, reinterpreting the Jewish heritage in light of environmental concerns and values (Bernstein 1997). The prophetic commitment to social justice led Arthur Waskow, the most important Jewish environmentalist, to promote a Jewish, religious, progressive environmentalism under the broad concept of "Eco-Kosher" (Waskow 1995, 1996).

In 1993, several Jewish environmental organizations coalesced into the Coalition on the Environment and Jewish Life. This umbrella organization has attempted to educate Jews about environmental matters, inspiring Jews to lead an environmentally sound communal life, beginning with the greening of synagogues, and calling Jews to lend their support to various legislative initatives and to enter inter-faith dialogue on environmental matters. Other small, non-profit organizations (e.g. Hazon, Teva, Isabella Freedman Learning Center, and Kanfei Nesharim) attempt to frame an environmentally aware Jewish way of living, but their impact on mainstream Jewish denominations is still very limited. The message that Judaism can be part of the solution to the ecological crisis, and that Judaism approaches the crisis in its own distinctive manner, has been slow to spread. Jews either assume that Judaism has little to say about environmental matters, or presuppose that Judaism and environmentalism are inherently incompatible.

There are historical and cultural reasons for this perception. The exile of the Jews from the Land of Israel has brought about major economic transformation in the life of the Jewish people. Urbanization of the ancient world, the occupational shift from agriculture to commerce in the Islamic world, and the restriction on Jewish ownership of land in the Christian world brought about growing alienation of Jews from their physical surroundings and a growing indifference toward the natural world. These historical conditions were exacerbated by the scholastic and bookish culture of Rabbinic Judaism that placed Torah study as the overarching ideal of Jewish religious life. Governed by study and ritual, Jewish religious life was text-centered rather than nature-oriented. For observant Jews in the pre-modern world, nature was good, but not holy; it could be made holy, or sanctified through the observance of divine commands that pointed to the divine Creator as the source of meaning and value, rather than to nature. Nature does not dictate what ought to be; only divine revelation articulates the moral norm. This is why Steven Schwarzschild spoke about the "unnatural Jew," (Schwarzschild 1984), a position that angered several Jewish environmentalists who missed the Kantian aspect of his claim. For Schwarzschild, as for many nineteenth-century Jewish theologians who were influenced by Kantian philosophy and were associated with the movement for religious reform, Judaism supersedes pagan religions. Whereas these traditions identify God with nature, Jewish monotheism expresses universal, rational morality that stands as a critique of nature. Unlike Spinoza, most modern Jewish thinkers regarded nature, especially as interpreted by Darwin's theory of evolution, to be amoral. Whereas nature has no room for the weak and the marginal, Judaism enjoins humanity to defend the weak and the socially marginal. In Judaism, what "ought" to be (morality) determines how humans relate to what "is" (nature). The Romantic infatuation with nature influenced modern Hebrew literature, especially Zionist literature, but not modern Jewish philosophy.

The Science of Ecology and the Theory of Evolution

Why has the science of ecology failed to inform Jewish reflections on the interaction between humanity and nature? One answer is found in the growing separation between Judaism and science during the modern period. The granting of civil rights to Jews in the nineteenth century enabled them to enter the universities that previously had excluded them. As part of the Jewish drive to integrate in European society and culture, Jews flocked to the natural sciences, especially the new disciplines of chemistry, microbiology, and medical biology (Charpa and Deichmann 2007: 5–36). In some scientific fields, Jews were disproportionately represented and reached outstanding achievements; other fields were shaped entirely or mainly by Jewish scientists. For Jews who wished to become part of European culture, modern science became a substitute for religion, although, ironically enough, it was the lifestyle of traditional Judaism, with its insistence on meticulous observance of laws and commitment to the pursuit of truth, that generated the personality type suitable for the rigors of scientific inquiry.

But no less important is the fact that the science of ecology, from its inception, was a Christian (and more specifically Protestant) discourse (Worster 1994; Cittadino 2006, Stoll 2006), a fact that Barrett and Jordan fail to mention. The science of ecology took shape in Lutheran countries (such as Germany and Denmark) and was planted in America in Congregationalist and Presbyterian denominations. For Protestant contributors to the science of ecology (such as John Muir), nature, which was viewed as pure, ongoing creation of God, served as the foil for human corruption. They all shared the view that the environment was an organism, and their holistic outlook was colored with mystical overtones. Since the science of ecology was con-figured within a Christian matrix, it is no wonder that Jews did not engage it, and instead chose to pursue natural sciences that were devoid of religious undertones (such as chemistry). Yet Jews did not ignore the core theory of the science of ecology—evolution. Informed of Darwin's variant of the theory of evolution by random selection, Jewish scholars were aware that natural selection challenges traditional Judaism. Only a handful of Jews engaged the scientific details of the theory of evolution; most focused on the *implications* of evolution for Judaism (Cantor and Swetlitz 2006). However, the debate was not merely theological; it also reflected diverse attitudes towards modernity, assimilation, and acculturation, pluralism within modern Judaism, and changing conceptions about the desired relationship between the Jewish minority and Christian majority.

In America, the Jewish engagement with Darwinian evolution came mainly after the publication of Darwin's *Descent* of Man in 1871 (Swetlitz 1999, 2006). A rigorous dispute erupted between leading Jewish theologians, who promoted the reform of Judaism. Rabbi Kaufmann Kohler endorsed Darwin's theory because he viewed it as a scientific proof for Reform Judaism: Judaism is a progressive religion that evolved over time; yet Kohler ignored Darwin's theory of natural selection. His opponent, Rabbi Isaac Mayer Wise, understood natural selection to be a natural law in which survival of the strongest prevails. Nature was indeed a "battle ground" which robbed the moral law, as taught by Judaism, of all its legitimacy. Rejecting Darwin, Wise developed his own theory about the history of life, one that acknowledged that the Jewish religion, like nature, is bound by the law of evolution, that is, progression from lower to higher forms. However, Wise did not adopt the theory of progressive revelation, and opposed the notion of graduate continuous transmutation of species. A few Reform rabbis followed Wise in rejecting evolution on the ground that it is improbable, but several more traditional rabbis among the reformers began to consider arguments against evolution on the basis of specific references in Genesis.

The debate between those who appealed to evolution to explain the progressive nature of Judaism, and the traditionalists who rejected evolution because it challenged Jewish observance, resulted in the emergence of a new Jewish denomination, Conservative Judaism. It would enable millions of Eastern European Jewish immigrants to America to modernize while remaining loyal to traditional Judaism. In the 1930s, Rabbi Mordecai Kaplan (1881–1983) elaborated the notion of Judaism as an evolving civilization, even though he did not accept key elements of Darwin's theory of evolution (Swetlitz 2006: 52–55). Similarly, Kaplan failed to draw the implications of the theory of evolution for belief in God, although he did develop a naturalistic world-view that viewed the universe as an organic totality. Ironically, when Kaplan articulated his new view of Judaism (Kaplan 1934), the science of ecology began to move away from its religious roots, with the rise of utopian visionaries who viewed ecology as the scientific underpinning of a new social order and the antidote to the excesses of modern civilization. In the mid-1930s, the term "ecosystems" replaced the term "ecology" in order to dissociate ecology from organic holism, which generated objectionable political implications on the political left (Communism) and on the political right (Fascism). After the Second World War, the study of ecosystems became increasingly mathematical, as concepts and metaphors from cybernetics and information science gave it a new flavor and direction (Stoll 2006: 64). Thus the *religious* "old ecology" was gradually replaced with *secular* "new ecology."

Ecological Strands in Modern Judaism

The distinction between "old ecology" and "new ecology" does not capture the nuanced reflections on the interplay between God, humanity, and nature in modern Jewish thought. Nonetheless, several Jewish thinkers can be labeled "ecological" because they highlight interaction and connectedness between all living and nonliving things; they reject the notion that nature has only instrumental value as a resource utility for human beings; and they promote a view of nature as a source of moral obligation and/or spiritual vitality. A few examples will illustrate ecological tendencies in Modern Orthodoxy, Zionism, and the Jewish Renewal Movement. The impetus for these ecological reflections did not come from the science of ecology, as some views held nature to be holistic, stable, and balanced, while others recognized nature as dynamic, perpetual flux.

Modern Orthodoxy

Modern Orthodoxy was founded in Germany by Samson Raphael Hirsch (1808–88), as a response to the movement for religious reform. While Hirsch endorsed the Jewish struggle for emancipation, and advocated openness toward European secular culture, he defended the traditional doctrine of "Torah from Heavens." He rejected the reformers' critique of rabbinic Judaism or their tendency to discard traditional practices and introduce new rituals. In continuation with rabbinic Judaism, Hirsch argued that God, the Creator of the world, gave humans the *right* to rule nature, but that this right came with the *duty* to treat God's created order in accordance to divine will. According to Hirsch, the human is part of nature, indeed a "brother to all creatures," but, like the first-born in a Jewish family, the human also has special privileges and obligation. Hirsch held that human mastery over nature is about the fulfillment of the human, rational, God-given free will, but the human right of mastery over nature depends on the extent to which the human will corresponds to God's will. The Torah itself thus discloses how nature is to be treated with justice and respect.

In his classification of Jewish laws, Hirsch placed the commandment "Do not destroy" at the head of the section on the *Hukkim*, which he defined as "statements concerning justice toward subordinate creatures by reason of the obedience due to god; that is, justice toward the earth, plants, and animals, or, if they have become assimilated to your own person, then, justice toward your own property, toward your own body and soul and spirit" (Hirsch 1969: 75). The *Hukkim* were legislated by God primarily for the protection of nature from human avarice and exploitation. Heedless destruction of nature reflects human arrogance and rebellion against God, but the laws of the Torah assure that humans behave wisely and judiciously to protect the integrity of the natural world and its perpetuation. For Hirsch, the created world exhibits not only inherent purpose, but also orderliness, intelligence, and interdependence. Given the remarkably wise design of the universe, humans have an obligation to protect the natural world. Indeed, Scripture prohibits copulation of diverse animals, grafting of diverse trees, yoking together of diverse animals, wearing of wool and flax, and mixing of milk and meat. These laws are rooted in the act of creation when God separated His creatures "each according to its kind." Through the Torah, the Creator of the world functions as the "Regulator of the world;" the human who was appointed as the "administrator" of God's estate executes the rules that ensure protection of nature. In Hirsch's analysis of the *Hukkim*, nature serves as a model for observance of divine commands and places its own demands or commandments on humans. Hirsch is an example of how Modern Orthodoxy can support Jewish religious environmentalism.

Zionism

Jewish religious environmentalism accords a place of honor to the Land of Israel, the Holy Land, whose well-being reflects the dynamic relationship between God and the People of Israel. The Bible makes clear that the Land of Israel was given to Israel conditionally: so long as Israel observes God's Will and follows His commandments, including the commandments for the treatment of nature, the land remains fertile, providing the People of Israel with abundance. But when Israel rebels against God and sins against God, the Land becomes infertile and desolate; when sinfulness abounds, God exiles Israel from the Land of Israel. Throughout their long exilic existence, Jews have yearned to return to the Land of Israel, but they postponed the fulfillment of this dream to the Messianic Age of the remote future. At the end of the nineteenth century, however, Zionism rebelled against traditional Jewish messianism by insisting that Jews leave their country of residence and settle in their ancestral home. Only in the Land of Israel could the Jews become a normal nation with its own political sovereignty, spoken language, and unique culture. For Zionist ideologues, the return of the Jews to the Land of Israel was not only a secularization of the messianic dream, it was also a call for the return of the Jews to nature in order to create a new "muscular Jew," a physically strong, fearless Jew who celebrates the rhythms of nature and derives vitality from nature.

The return to nature played a central role in the original and highly creative thought of Aaron David Gordon (1856–1922), the spiritual leader of Labor Zionism (Schweid 1985: 157–70). Although Gordon grew up in an Orthodox home in Russia, on his own he studied European philosophy and literature, and was particularly influenced by Friedrich Nietzsche, Henri Bergson, and Leo Tolstoy. Settling in Palestine in 1904, Gordon joined the agricultural settlements and experienced the hardship of pioneers' life in order to exemplify his call not just for the Jews, but for all of humanity. For Gordon, the Jews' return to their land symbolized humanity's return to nature and the renewal of the relationship (indeed the covenant) between humanity and nature which was destroyed by culture, especially modern, urban, industrialized culture.

Gordon contended that human beings stand outside nature by developing three major postures: as artists, humans express love of nature's beauty; as scientists, they explore nature's mysteries; and as engineers, they use human technology to exploit nature's resources. All three postures are mistaken because they deny that humans are part of nature. Instead, humans should strive "to live with and in nature," because the universe is an organic totality enlivened by divine energy that pulsates through all levels of reality. Emerging out of the organic totality, humans have reached the most developed state of being: culture and self-consciousness. Yet these are also the causes of human alienation from nature. The human tragedy is especially severe in the case of Jews because they seek to assimilate into Western Christian culture, denying their roots in the Land of Israel, the Jewish nation, and the Hebrew language. Only agriculture and farming could reconnect modern, alienated Jews to the sources of cosmic creativity, enabling them to live most authentically as Jews as well as human beings, an ideal he exemplified in his own life.

A. D. Gordon's thought has a strongly pantheistic character, which could be traced to various intellectual sources, both Jewish (Kabbalah and Hasidism) and non-Jewish (Bergson and Tolstoy). He viewed nature as an organic totality out of which emerges human consciousness, and he called the totality of nature *havaya*. Literally, this means "being," but it is in fact the four letters of the Tetragrammaton arranged in a different order. Secularizing the notion of divine immanence, Gordon understood Being as a dynamic, ever-changing, living force in constant flux that pulsates throughout all levels of reality. Humans can experience the hidden aspect of nature only through direct intuition, but such experience can be attained only by means of manual labor that overcomes the alienation of mankind from nature. For Gordon, the regeneration of humanity and the regeneration of the Jewish people could come only through the development of a new understanding of labor as the source of genuine joy and creativity. Gordon's holistic views are susceptible to the critique of "old ecology" put forth by Barrett and Jordan, and others (Sideris 2003), but this critique cannot address the religious assumptions of Gordon's world-view.

Jewish Renewal Movement

The Zionist call for the return of the Jews to nature illustrates the traditional Jewish *idea* of teshuvah. The term is usually translated as "repentance," but its Hebrew stem connotes both "to return" and "to reply," so that *teshuvah* means both a movement of return to one's source, to the original paradigm, and simultaneously a response to a divine call. *Teshuvah* thus involves a spiritual reorientation or a transformation of the self away from an inauthentic way of life, lacking creative force, toward the authenticity that is characterized by continual creativity and renewal. In America during the second half of the twentieth century, this concept gave rise to the Jewish Renewal Movement.

The catalyst for it was Rabbi Abraham Joshua Heschel (1907–72), a Polish Jew who was born into a Hasidic family, but who also received modern university training. Heschel managed to flee Nazi-occupied Poland and settled in 1941 in America, where he inspired scores of alienated American Jews to find their way back to the sources of Judaism in order to heal the atrocities of modernity, which culminated in the Holocaust. Heschel's ecologically sensitive Depth Theology spoke of God's glory as pervading nature, leading humans to radical amazement and wonder; viewed humans as members of the cosmic community; and emphasized humility as the desired posture toward the natural world (Kaplan 1996). Another product of Hasidism, and a refugee from Nazi-occupied Europe, was Zalman Schachter-Shalomi (b. 1924), who founded the Jewish Renewal Movement in the late 1960s. His creative reinterpretation of Judaism combined a "Gaia consciousness" with psychological interpretation of Lurianic Kabbalah and New Age spirituality. Schachter-Shalomi urged a paradigm shift from monotheism to pantheism within Judaism, but did not intend to revive neo-pagan pantheism. Instead, he offered contemporary Jews a new way to infuse Jewish life with rituals that envision "a God who is an integral part of all human civilization and all of humanity has a specific responsibility for that relationship" (Magid 2006: 65). Unlike the rituals discussed by Barrett and Jordan, the rituals of the Jewish Renewal Movement are all contemporary variants of traditional Jewish practices (for example, the pilgrimage festivals). Schachter-Shalomi's ideas were systematized into Jewish ecology by Arthur Green (b. 1941), who, under the influence of Kabbalah, presented a holistic view of reality in which all existents are in some way an expression of God and are to some extent intrinsically related to each other (Green 1992, 2003). From the privileged position of the human, Green derives an ethics of responsibility toward all creatures that acknowledges the differences between diverse creatures, while insisting on the need to defend the legitimate place in the world of even the weakest and most threatened of creatures. For Green, a Jewish ecological ethics must be a set of laws and instructions that truly enhances life. Green's ecotheology does not pay sufficient attention to the "dark" of nature, and does not explain how human conduct, guided by divinely revealed law, may actually address it. Thus Green, like Gordon and Heschel, is open to the criticism leveled by Barrett and Jordan.

Ecology and the Jewish Imperative of Responsibility

To some extent, the Jewish Renewal Movement and all late-twentieth century Jewish ecological thinking can be viewed as a belated response to the catastrophe of the Holocaust, a determination of the Jewish people to renew themselves so as "not to give Hitler a posthumous victory," to use the famous formulation of Emil Fackenheim. The Nazis' attempt to exterminate the Jewish People because of their regard for the Jews as sub-human was the most distorted application of evolutionary theories. Treating Jews as vermin, the Nazis applied Zyklon-B gas with the aim of totally eradicating them, using bureaucratic efficiency and the most advanced science and technology for demonic purposes. The struggle between Nazism and Judaism raises some poignant questions about paganism as a world-view that does not allow for the possibility of transcendence, and that takes the world of the senses to be ultimate reality. The horrendous results of Nazism for Jews remind us that Nazism was the most consistent assault on the Jewish notion that physical environment is not inherently sacred, and that only when humans act in accord with divine commands can nature become holy. While nature can be a source of spiritual inspiration, it is important to remember that nature is also violent, competitive, ruthless, and destructive, precisely as the "new ecology" teaches. Nature, indeed, does not care about the sick, the weak, and the deformed; it disposes of them in the relentless struggle for survival. Nature does not establish moral values that can create a just society in which the needs of the sick and the poor are addressed. These moral values, which constitute the Jewish ethics of responsibility, are revealed by God and implemented by humans who wish to sanctify nature as they strive to become like God.

No-one understood the lessons of the Holocaust better than Hans Jonas (1903–93), a German Jewish Zionist who fled Nazi Germany in 1934 and settled in Palestine in 1935 (Wiese 2007). During the Second World War, Jonas volunteered with the Jewish Brigade of the British army, which brought him back to Germany as a victor. Yet the war experience made him shift to philosophy of biology, or more precisely to the philosophy of organic life. Jonas' philosophic response to the total destruction practiced during the Second World War was to invest matter itself with *inherent moral meaning*, presenting the human as the final manifestation of nature's "needful freedom" (Jonas 1966, 1984, 1996). His philosophy of organism starts with the phenomenon of metabolism, and interprets organic life and individual organisms in terms, concepts, and categories that transcend Cartesian dualism, idealism, and physicalist materialism. For Jonas, organic life itself is an ontological revolution in the history of matter, a radical change in matter's mode of being. By giving organisms on all levels their philosophical and moral due, animate nature was philosophically rehabilitated. The radical split between nature and ethics, between "is" and "ought," was thereby bridged. Jonas asserted that value and disvalue are not human constructs, but "essential to life itself [since] every living thing has a share in life's needful freedom." Jonas did not appeal to revelation, even when he reworked traditional Jewish concepts such as creation in the "image of God." Yet Jonas' insights were profoundly Jewish; they resonate with traditional Jewish moral passion and the prophetic call to protect the needy and powerless. Jonas was less stunned by the innumerable material forms and processes of life than by the very fact of life itself, and especially by organic *life's capacity for moral responsibility, evidenced in human beings*. The very fact that, in a vast universe characterized largely by inorganic, dead matter, there has emerged animal and moral being as a revolt against death and valuelessness led Jonas to enjoin us to protect organic life into the indefinite future. Jonas is the best proof that the particular historical circumstances of Jews could inspire an environmental philosophy for humanity that attempts to cope with an ecological crisis of its own making. Since human activities are largely, although not exclusively, responsible for the current ecological crisis, humans have the *responsibility* to mend the world that belongs to God, precisely as traditional Judaism teaches.

In contrast to the Jewish emphasis on responsibility, Barrett and Jordan maintain that the only way to come to terms with the truths of the science of ecology is for humans to acquire a sense of shame. The authors invoke Jean Paul Sartre, who held that the very existence of an onlooker implies shame. However, Sartre did not explain shame; he only explained it away. What is lacking in Sartre's analysis of shame, as Christian critics have already noted, is an adequate accounting for the sense of culpability that is inseparable from shame: if the Other is merely an intruder, we would only feel annoyed and attempt to eliminate the intruder. But the glance of the Other does not generate shame unless it falls upon some sin that exists objectively within the shamed one. Barrett and Jordan consider the traditional Christian notion of guilt and sinfulness a "tragic mistake" in our framing of human relationship to nature, but the horrific experience of Jews in the Holocaust attests that it is much more tragic when human beings dispense with notions of guilt, culpability, and sinfulness occasioned by one's harming of the Other. After all, these are the values that Nazi ideology and culture conspicuously lacked, leading the Nazis to commit unspeakable atrocities. Sartre's concept of existential shame is not a promising point of departure for restorative ethics. Biblically rooted Christian eco-theologies are more in accord with the Judaic imperative of responsibility than the secular views promoted by Barrett and Jordan.

We must never abandon this responsibility, even though we will continue to debate how to translate the idea of responsibility into sound environmental policies in light of the science of ecology.

References

Bernstein, Ellen (ed.) (1997) *Ecology & The Jewish Spirit: Where Nature and the Sacred Meet*, Woodstock, VT: Jewish Lights Publishing.

Cantor, Geoffrey and Marc Swetlitz (eds) (2006) *Jewish Tradition and the Challenge of Darwinism*, Chicago, IL: University of Chicago Press.

Charpa, Ulrich and Ute Deichmann (2007) *Jews and Science in German Contexts*, Tübingen: Mohr Siebeck.

Green, Arthur (1992) *Seek My Face, Speak My Name*, Northvale, NJ: Aronson.

——(2003) *Ehyeh: A Kabbalah for Tomorrow*, Woodstock, VT: Jewish Lights Publishing. Hirsch, Samson Raphael (1969) *The Nineteen Letters on Judaism*, New York: Feldheim.

Jonas, Hans (1966) *The Phenomenon of Life: Toward a Philosophical Biology*, New York: Harper & Row.

——(1984) *The Imperative of Responsibility: In Search of an Ethics for the Technological Age*, Chicago, IL: University of Chicago Press.

——(1996) *Mortality and Morality: A Search for the Good after Auschwitz*, Lawrence Vogel (ed.), Chicago, IL: Northwestern University Press.

Kaplan, Edward K. (1996) *Holiness in Words: Abraham Joshua Heschel's Poetics of Piety*, Albany, NY: SUNY Press.

Magid, Shaul (2006) "Jewish Renewal, American Spirituality, and Post-Monotheistic Theology," *Tikkun* (May/June): 62–66.

Rosenblum, Noah H. (1976) *Tradition in an Age of Reform: The Religious Philosophy of Samson Raphael Hirsch*, Philadelphia, PA: Jewish Publication Society.

Schwarzschild, Steven S. (1984) "The Unnatural Jew," *Environmental Ethics* 6: 347–62; reprinted in Yaffe, Martin D. (2002) *Judaism and Environmental Ethics: A Reader*, Lanham, MD: Lexington Books, 267–82.

Schweid, Eliezer (1970) *The World of A.D. Gordon* (in Hebrew), Tel Aviv: Am Oved.

Sideris, Lisa (2003) *Environmental Ethics, Ecological Theology and Natural Selection*, New York: Columbia University Press.

Stoll, Mark (1999) "American Jewish Responses to Darwin and Evolutionary Theory, 1860–90," in *Disseminating Darwinism: The Role of Place, Race, Religion and Gender*, Ronald L. Numbers and John Stenhouse (eds), Cambridge: Cambridge University Press, 209–46.

——(2006a) "Creating Ecology: Protestants and the Moral Community of Creation" in *Religion and the New Ecology: Environmental Responsibility in a World in Flux*, David M. Lodge and Christopher Hamlin (eds), Notre Dame, IN: University of Notre Dame Press, 53–72.

——(2006b) "Responses to Evolution by Reform, Conservative and Reconstructionist Rabbis in Twentieth-Century America," in *Jewish Tradition and the Challenge of Darwinism*, Geoffrey Cantor and Marc Swetlitz (eds), Chicago, IL: University of Chicago Press, 47–70.

Tirosh-Samuelson, Hava (2006) "Judaism," in *The Oxford Handbook of Religion and Ecology*, Roger S. Gottlieb (ed.), Oxford and New York: Oxford University Press, 25–64.

Waskow, Arthur (1995) *Down-to-Earth Judaism: Food, Money, Sex and the Rest of Life*, New York: W. Morrow.

——(1996) "What Is Eco-Kosher?," in *This Sacred Earth: Religion, Nature, Environment*, Roger S. Gottlieb (ed.), New York: Routledge, 297–300.

——(ed.) (2002) *Torah of the Earth: Exploring 4,000 Years of Ecology in Jewish Thought*, Woodstock, VT: Jewish Lights Publishing.

Worster, Donald (1994) Nature's Economy: *A History of Ecological Ideas*, 2nd edn, Cambridge: Cambridge University Press.

Yaffe, Martin D. (ed.) (2002) *Judaism and Environmental Ethics: A Reader*, Lanham, MD: Lexington Books.

DISCUSSION QUESTIONS

1. What gave rise to the Jewish environmentalist movement? What was the movement reacting to, and what was the Coalition on the Environment and Jewish Life?
2. How was evolution tied to the emergence of the Conservative Jewish denomination?
3. Explain the doctrine referred to as the "Torah from Heaven," and give examples to explain your points.
4. Explain Aaron David Gordon's beliefs on the relationship between humans and nature.
5. Describe what the word *ecotheology* could mean, and highlight examples.
6. According to the article, what were the life lessons for Hans Jonas?

POLITICS

Israel is a democratic country, and the Knesset is the legislative branch of its government. The Knesset appoints the president and prime minister, and it can pass laws with a simple majority. It also oversees any government function through a cabinet and committees. However, since the creation of the State of Israel, trying to balance a democracy while remaining a Jewish nation has proved difficult. The following article explores the initial tensions between Judaism and democracy within Israel.

ARTICLE 2

THE TRANSFORMATION OF JUDAISM IN ISRAEL

By Robert D. Lee

It matters who is a Jew in Israel. The approximately 23 percent of Israel's legal citizens, roughly 1.5 million people, who are not considered Jews under Israeli law, most of them Arabs, do not enjoy full citizenship rights in the state of Israel. While the Declaration of the Establishment of the State of Israel promises non-Jews "full citizenship," the religious symbols of these "minority" populations do not appear in Israel's institutions. Although minority religions receive funding from the state, the monies come from a fund separate from the one that subsidizes Jewish (Orthodox) institutions. Furthermore, the Israel Land Authority controls 93 percent of the land within the pre-1967 boundaries of Israel and decides who is entitled to lease property for which purposes.[27] Serving the needs of Jewish immigrants has been a primary purpose of public policy since the founding of the Jewish National Fund, a quasi-governmental NGO, in 1901.

Recent developments suggest some willingness to lessen the privileges of Jewish identity. The Supreme Court decision in the Qaadan case of 2000 challenged the right of the Jewish National Fund to discriminate between Jews and non-Jews in leasing land, and amendments to the Law of Return extended rights at least as far as relatives of Jews. These changes constitute an ideal of citizenship more in line with liberal-democratic ideology. The recent activism of the Supreme Court, coupled with the Knesset's adoption of two basic laws—Human Dignity and Liberty, and Freedom of Occupation—constitutes a gradual shift toward a value system that prizes equal application of the law over Jewish collective objectives. However, both the Law of Return and the Israel land laws grant Jews preferential status as citizens of the state of Israel.

Israel's Jewish identity has proved stable in its inconsistency. On the one hand, the state cannot renounce its Jewishness without undermining its reason for being. Hard as Zionists might try, they did not succeed in defining a Jewish people without some reference to religion. Beliefs, scholarship, rabbinic leadership, common rituals and ceremonies—all these held Jewish communities together in the Diaspora. An Israel without Jewishness cuts itself off from the Diaspora and from its links with Jewish history. On the other hand, Israel cannot adopt unmitigated Orthodox definitions of Jewishness without cutting itself off from the nonorthodox Diaspora and from important elements of its own population, such as the Russian immigrants. Even a watered-down commitment to Jewishness makes it difficult if not impossible for Israel to achieve the democratic commitment contained in the Declaration of Independence. Non-Jews lack full status in the state. Arabs, who are Christians and Muslims, are third-class citizens behind the Jews of European origin (Ashkenazim), who enjoy primacy in the society, and the Mizrachim, who are still struggling for parity with the Europeans.

The conflict between Israelis and Palestinians may help stabilize the inconsistencies inherent in Israeli identity. The unexpected victory of 1967 opened the way for a new religious Zionism committed to settle the West Bank and Gaza—lands linked to Jewish history. These initiatives tipped Israeli identity toward the religious. The ratio of Israelis who put their Jewishness ahead of their identity as Israelis increased slightly and became a majority. The settlers and their supporters saw themselves as Jews doing the work of God. They attacked "mere" Israelis, who wanted to trade land for peace with the Arab states and the Palestinians. More generally this conflict reinforces Israelis' self-perceptions as a persecuted minority. Until this conflict ends, it seems unlikely Israeli society will embrace full democratic inclusion of its non-Jewish citizens.

Israel's identity remains distinctly Jewish. The Jewishness of the state results from a set of political decisions that reflect divergent and even contradictory notions of Judaism and Jewishness. The state remains too Jewish for some and insufficiently Jewish for others. The privileges accorded Jewish identity make the goal of full and equal citizenship for non-Jews unattainable. The elaborate use of religious symbols to legitimate the secular state strike some as a travesty. Yet many religious Zionists see the state as halfhearted and even hypocritical in its commitment to its sacred mission, the realization of God's promise to the Jews. Ultimately the Jewishness of Israel reflects a set of political decisions taken to realize the ambitions of Zionism. Nationalism requires an identity for a people, and no one has yet devised a definition of the Jewish people that does not refer to religion. The religious identity of the state is not a result of religious imperatives but of nationalist necessity.

Ideology

Israel's Declaration of Independence frames one of the central debates over Israeli political ideology. Ratified in 1948, it promises that the Israeli state will be "based on freedom, justice, and peace as envisaged by the prophets of Israel." While the document recognizes the visions of the Jewish prophets, it denies political authority to rabbis and makes no claim that the law of the state will reflect or abide by the traditional rabbinical laws of the Jewish religion. Instead the declaration explicitly promises that elected authorities will govern the state of Israel in accordance with a constitution written by an elected constituent assembly. How much and in what ways should the law of the Jewish state reflect the religious law of the Jewish tradition? Should the Jewish state be governed exclusively by elected politicians and bureaucrats? Or should the traditional leaders of the Jewish faith who serve as interpreters of divine law hold positions of authority?

Pre-State Discourse

The most powerful politicians of the Zionist movement had no interest in turning over government power to rabbinical leaders. The religious conceptions of these men and women reflected the socialist and democratic ideals of the modern Zionist movement. The founders framed their call for the Jewish state in terms of political necessity, as had Theodor Herzl. They saw themselves as champions of a homeless people dispossessed from its native land, not of a movement to create a religious state. David Ben-Gurion, Israel's first prime minister, assured the devout that the government would show "understanding of your religious feelings and those of others," and would be "devoted to the values of Judaism" but would not make Israel into a theocratic state or impose religious law upon the people.[28] In the constitutional debate that followed the ratification of the declaration, a member of the Knesset loyal to Ben-Gurion proclaimed: "As a socialist and an atheist, I could never endorse a program that included a religious model."[29]

Few ultraorthodox Jews of the pre-state period regarded the Zionist project as divine. While Orthodox Jews did regard the Land of Israel (Eretz Yisrael) as sacred, most did not attribute holiness to the state of Israel (Midinat Yisrael), whose architects largely considered themselves secular Jews. Consequently, few Orthodox Jews believed in a divine mandate for the creation of a halakhic state and were thus willing to accept compromise on religious matters.

In the early nineteenth century, much of the Orthodox establishment evoked a rabbinical tradition that discouraged the formation of a Jewish political entity in the land of Palestine. Although Orthodox Jews everywhere believed that the "End of Days" and the coming of the Messiah would eventually bring the dispersed Jewish communities back to their homeland, most Orthodox rabbis taught that this migration would be of a miraculous nature. The Mizrachi movement among Orthodox Jews began a cautious shift in this long-held tradition. Created in Vilna in 1902 among Orthodox delegates to the Zionist Congress, Mizrachi sought to foster cooperation between two separate responses to modernism: Orthodoxy and Zionism.[30] In 1912, Agudat Israel split with Mizrachi to engage in Jewish settlement of Palestine outside the auspices of the World Zionist

Organization (WZO). Both groups regarded the secular Zionist movement with suspicion but came to cooperate with the WZO on a practical level during the 1940s, when European persecution of Jews reached catastrophic proportions.

Suspicion of Labor Zionism led a large portion of deeply Orthodox Jews in Palestine in 1948 to reject the formation of a halakhic state and to deny that the modern Israeli state was holy or theologically significant. These groups feared that an Israeli state claiming a basis in modern Judaism would undermine the Orthodox cultural establishment in Israel. They attempted to persuade the Labor Zionists to refrain from engaging in Jewish cultural activities.

In the early twentieth century, however, there emerged an alternative Orthodox discourse that invested the Jewish state with religious meaning. Rav Abraham Isaac Kook argued that the establishment of a Jewish state in the Land of Israel was an "advent of redemption" and had not only religious but also messianic implications.[31] Yet he was willing to accept that Israel would at least temporarily be a secular, nontheocratic state. While he hoped and anticipated that Jews who had rejected rabbinic authority and traditional law would repent in the final stage of messianic redemption, he welcomed the work of the Labor Zionists in taking the land and creating a space for the physical salvation of the Jewish nation. These were necessary conditions for the spiritual salvation he believed would come later. Kook helped to create a discourse that permitted the formation of a sacred but nonhalakhic Israeli state. These Orthodox discourses eased the burden of the early Zionist leaders, who feared and rejected the idea of a halakhic state. Significant portions of the Orthodox population of Palestine and important Orthodox leaders discouraged involvement in the Jewish state, lest the integrity of traditional Judaism be compromised. Other Orthodox Jews and leaders accepted the sanctity of the state of Israel without demanding that it be governed immediately by religious law or religious leaders. Consequently, no significant movement for theocratic government surfaced as the Jewish community in Palestine (the Yishuv) prepared for statehood.

The Founding

Lack of demand for theocracy did not, however, prevent fierce debate over the role of halakhic law in Israel and the allocation of governmental power to Israel's religious leaders. Those questions dominated debate in the constituent assembly, which quickly declared itself the sovereign parliament, the Knesset, and eventually prevented the adoption of a constitution. One member representing Agudat Israel argued that "only the Torah Law and tradition are sovereign in the life of Israel."[32] While the religious forces recognized that the creation of a halakhic state would be impossible, they fiercely opposed the drafting of a secular constitution, which they feared might sever the tie to the past.[33] "If the time is not yet ripe for our constitution to be based on the laws of our Torah," argued Agudat Israel, "it is better that no constitution be passed and that we not be untrue."[34] The Knesset quit trying to draft a constitution, but the ideological battles over the role of religion in government were far from over.

Israeli leaders used two strategies of tempering ideological debates and creating a degree of ideological stability. Their first strategy was to frame policy and law so that they conformed to the basic tenets of Jewish tradition even if they did not precisely reflect the halakha. Ben-Gurion used this strategy frequently, as demonstrated in one of his speeches: "Our activities and policy are guided not by economic considerations alone but by a political and social vision that we have inherited from our prophets and imbibed from the heritage of our greatest sages and the teachers of our own day."[35] Politicians such as Ben-Gurion embraced the political and social vision of Jewish religious leaders in an attempt to legitimate modern interpretations of ancient principles and traditions.

This rhetorical strategy required a clever ability to avoid discussing religious legitimacy with the Orthodox rabbinical establishment. Ben-Gurion insisted that the "Rock of Israel (*tsur* Israel)" was to be found in the "State of Israel and in the Book of Books."[36] And on at least one occasion, he referred to the modern Israeli state as the Third Kingdom, comparing it to the ancient Jewish communities of biblical times. By focusing on the Bible and on ancient Israeli civilization, Ben-Gurion attempted to steer ideological debate away from the Talmud and the halakha. This strategy permitted him to argue for the sanctity of state policies without deferring to the authority of contemporary religious leaders, who claimed to be the rightful interpreters of religious law.

Ben-Gurion's approach amounted to more than rhetoric. Attempts to embrace the basic principles of Judaism, and to couch legislation and institutions in the language of religious tradition, influenced the character of Israeli institutions and law. A letter Ben-Gurion wrote to the ultraorthodox Agudat Israel has long served as the foundation for political debate over religious issues in Israel. In this letter, which came to be known as the status quo letter, Ben-Gurion promised that the religious character of the Jewish state would be respected and reflected in the policies and practices of the Israeli government. The first two points of the letter reflect the strategy of Israeli political leaders, described above. "Saturday shall be the national day of rest," the letter reads, and "the laws of *kashrut* will be observed in all government kitchens." These promises did not guarantee observance of halakha, but committed the state to honor Jewish tradition in some domains.

To implement the promises of the status quo letter, the government passed the Law of Working Hours and Rest in 1951, limiting the workdays of Jewish employees and effectively barring most industrial activity on the Sabbath. Framed in modern legal language, the law did not, however, apply to self-employed Jews or to government utilities and services crucial to the proper functioning of the state. While the legislation honored a principle, it did not adhere to the letter of the halakha. Neither did it interfere with the practical necessities of the state.

A similar strategy led to compromise on kashrut. While the Israeli government was not willing to pressure its citizens to maintain these dietary laws in their own homes, the government did maintain kashrut in its kitchens and other facilities. Positive law prevented state-owned companies from importing pork and any other meat not slaughtered in accordance with religious law. The state hired Orthodox inspectors to ensure that military food was prepared according to religious standards. These and other compromises served to quell ideological conflict by covering state institutions with a veneer of religion. They went far enough to secure the continued participation of Orthodox Jews in government institutions, and especially in the military, without subordinating government authority to religious leadership.

The strategy of compromise proved insufficient. Traditional religious Jews, and the rabbinical establishment of Palestine, rejected Labor Zionism's efforts at modernizing Jewish tradition to fit a nationalist agenda. While many Orthodox Jews accepted compromise over the expression of religious ideals in the state's public activities, they would make no concessions in other domains. In response, Israeli leaders used a second strategy to soften ideological debates. The status quo letter promised that religious organizations would continue to receive funds from the state to maintain their autonomous school systems and that religious courts would have exclusive authority over matters of personal status, including marriage, divorce, conversion, and burial.

Israeli citizens considered Jewish cannot engage in civil marriage within the state of Israel. Although the state recognizes civil marriages performed outside Israel, government employees appointed by religious courts of the fourteen religious denominations recognized by the Israeli government issue marriage certificates.[37] Judaism is but one of the fourteen, albeit the largest. These religious courts grant marriage licenses and perform ceremonies according to standards of religious law upheld by religious authorities in Israel. Governmentally recognized Jewish leaders supervise all conversions to Judaism and manage funerals and Jewish cemeteries.

In addition, the status quo agreement promises that "full autonomy of every education system will be guaranteed." At the state's inception, the Ministry of Education began to allocate funding among its many autonomous school systems according to the size of the populations they served. This includes explicitly Muslim, Christian, Druze, Bahai, and Jewish schools; all the preceding schools are free to teach religious curricula to their students as they see fit. By permitting autonomy, secular elites sought to dissuade religious groups from seeking influence over a broad spectrum of the state's laws and functions in return for domination of significant but limited spheres of policy. While government officials handle a broad range of state functions according to codes of civil law, religious officials administer other domains according to their own interpretations of sacred texts and tradition.

Early political leaders employed two major strategies to deal with the perplexing paradoxes of a Jewish republic. First, they attempted to connect the law and institutions of the Israeli state apparatus to a modified version of Judaism. Second, they tried to grant specific spheres of influence to religious leaders and some autonomy to religious institutions. While protests over religious legislation did occasionally erupt, the early Israeli leadership managed to maintain a balance. Many religious leaders expressed discontent from time to time but also often articulated their gratitude to Ben-Gurion and his allies for their cooperation. Menahem Parosh of Agudat Israel said: "Ben-Gurion gave us more than anyone else, because he understood that if the state did not make concessions to us we would have to leave the country, and this he did not want."[38] As long as their own spheres of authority were not challenged or disturbed, Orthodox leaders could accept that the state of Israel did not abide by halakha or recognize their authority on other matters. And although some citizens rebelled fiercely when they thought religious leaders and authority figures had overstepped boundaries, few citizens were willing to fight for the total disenfranchisement of these leaders.

Ideological Shift

While there were occasional challenges to the status quo agreement, its basic lines prevailed through the first twenty turbulent years of Israeli history. Orthodox leaders had no practical ideas about how a halakhic state might be governed.[39] Because a modern nation-state must operate utilities, ensure security, and maintain other services, the Israeli government could not observe the halakhic prescriptions of the Sabbath. Conversely, separation of religion and state did not seem possible or desirable. Most Israeli Jews, even those who consider themselves secular, wanted (and still want) the state to maintain its Jewish character for nationalistic reasons. Most Israeli Jews accepted the laws of the state of Israel as legitimate.[40]

In the 1960s and 1970s, however, ideological shifts in the Israeli political landscape eroded support for the fragile balance created by the status quo agreement. First, Israel absorbed more immigrants in comparison to its population than had any other country.[41] Many of these immigrant groups belonged to religious communities. They did not share the ideals of the secular Zionists or support the idea of modernizing the Jewish faith to fit contemporary needs. Government mistreatment of the immigrants bred discontent, and attempts to indoctrinate them through education fomented hostility toward the ruling coalition. Second, two deeply traumatic events, the Six-Day War of 1967 and the Yom Kippur War of 1973, contributed to ideological reconfiguration. Although the stunning victory in 1967 left many with a feeling of elation and national pride, it ended with the Israeli occupation of vast territories and Arab populations that posed new and difficult questions. The more costly 1973 war had a deeply traumatic effect on the Israeli people and divided previously unified coalitions. It led to diplomatic negotiation and eventual peace with Egypt, but Israeli sentiment remained deeply divided about giving up more territory to create a Palestinian state.

The wars of 1967 and 1973 sparked shifts in Israel's most Orthodox communities. Rav Kook's son, Rabbi Zvi Yehuda Kook, interpreted the two battles and the acquisition of the historically and religiously significant areas of the West Bank as indicative of messianic times.[42] Through the institution of Merkaz Harav, a highly influential religious academy established by his father, the rabbi began to articulate religious arguments for expanding the Israeli state and settling the occupied territories. While the senior Kook had argued that redemption of the land was a precursor to the spiritual redemption of the Jewish people and the world, Zvi Yehuda Kook portrayed the conquest of the land as a holy endeavor in and of itself.

A new generation of religious leaders began to utilize the vocabulary of Jewish tradition and to formulate a new ideology that justified and even commanded the forceful conquest of the Land of Israel. Rabbis such as Moshe Rom interpreted wars of conquest as "mandatory in Jewish tradition and argued against ceding any land to non-Jews as a religious sin."[43] Moshe Moskowitz, a leader of Agudat Israel, declared, "Do not doubt that the land and the people will find their mutual redemption."[44] These militant religious Zionists spread their ideas through a large network of yeshivas, academies of Jewish learning, where they found audiences of thousands of young, idealistic Jews.

The new ideology sparked settlement, and settlement solidified ideology. In 1968 Rabbi Moshe Levinger moved with a small group of pious followers to the city of Hebron. As one of four Jewish holy cities and the site of the cave of *machpelah*, thought to be the burial place of three of the Jewish patriarchs and their wives, Hebron held enormous religious significance to religious Jews.

But the city was inhabited almost exclusively by Arabs, most of them Muslims who, like Jews, honor Abraham as the founder of their religion. Levinger began a settlement called Hameuchad, often considered the first religious settlement established in the occupied territories. A short time later, religious settlers in Hebron organized the Gush Emunim (Bloc of the Faithful) around the militant religious ideology of Zvi Yehuda Kook. These settlers saw conquest of the entire biblical land of Israel as the divine right and obligation of the Israeli state; they claimed that this divine mandate transcended positive law. "We are commanded by the Torah according to the will of God," proclaimed one official Gush Emunim publication, "and therefore we cannot be subject to the standard laws of democracy."[45]

Rabbi Meir Kahane, who moved from the United States to Israel in 1971, took the new ideology of militant religious Zionism a step further by explicitly justifying the use of violence to secure the whole Land of Israel. Kahane not only hailed conquest as a noble and sacred activity, but also claimed that militant conquest was the primary, divinely inspired purpose of the Israeli state. In an essay written in 1976, Kahane declared, "The State of Israel was established not because the Jew deserved it, for the Jew is as he has been before, rejecting God, deviating from his paths and ignoring his Torah." Kahane argued rather that the divine purpose of the state was "a Jewish fist in the face of an astonished Gentile world that had not seen it for two millennia."[46] Kahane proclaimed that "when the Jew is beaten, God is profaned!" He preached that the state of Israel was God's instrument of holy war.

Kahane expanded his ideological formulation into a full-fledged apocalyptic prophecy. In 1980, after being arrested by Israeli police for his involvement in a plot to blow up the Dome of the Rock, Kahane claimed that on the basis of biblical texts, the formation of the Jewish state in 1948 was a sign of the impending messianic redemption. According to Kahane, the forty years following the establishment of the state were a grace period in which Jews could choose to repent their sins, obey God's law, and experience "a great and glorious redemption," or choose to remain in defiance of God and face a redemption brought about through "terrible sufferings and needless agonies; . . . holocaust more horrible than anything we have yet endured."[47] Predicting an impending catastrophe, Kahane advocated a "truly Jewish state, not a Hebrew-speaking gentilized one."[48] He argued, among other things, that democracy should be suspended in Israel, that Arabs should be stripped of political rights and excluded from all spheres of work, that Jewish sovereignty should be proclaimed over all of the Land of Israel "by virtue of the promise of the Almighty," that Israeli public schools should teach a fully religious curriculum, and that dietary laws, censorship laws, and dress codes should be strictly enforced in public venues according to halakha.[49]

While Kahane's religious ideology and apocalyptic predictions galvanized his inner circle, the rabbi also managed to appeal to a larger audience by couching his ideas in more secular and pragmatic language. He warned Israelis that the demographics of Israel were the real danger, playing on nationalist anxiety felt throughout Israeli Jewish society that a Jewish majority would no longer exist in Israel.[50] He appealed to poor Mizrachim by claiming that cheap Arab workers were undermining the Israeli labor market, and he utilized racist biases to garner support. After several electoral bids to gain a seat in the Knesset, Kahane's party, Kach, gained 1.2 percent of the vote in 1984, enough to win a seat in parliament, but the Knesset barred Kach from taking its prize.[51]

The ideology of the settler movement developed on the fringe of the Israeli political spectrum, but the religious parties used the movement's precepts and spirit to mobilize political coalitions and push for favorable policies. In 1977, the defection of the National Religious Party after a series of disputes over Sabbath regulations brought down a Labor government under the leadership of Yitzhak Rabin.[52] The right-wing, nationalist coalition that took power in 1977 shared many common interests with religious settlers, although it never adopted the ideological discourse of the militant settler movement. The religious parties exacted a high price for their support of the coalition in terms of increased state support of religious institutions and religious education.

The compromises and strategies derived from the status quo agreement did not work in the new political circumstances. While the Orthodox community had been primarily concerned with preserving its own privileges in the Israeli establishment, the ideology of militant religious Zionism demanded radical action. Because the radicals believed that the conquest of the land was a divine duty, they were unwilling to dismiss the actions of the Israeli state as theologically inconsequential, as had non-Zionist Orthodoxy. And believing that they were living in times of impending messianic redemption, militant religious Zionists were unwilling to agree with the elder Kook that Israel could wait to fulfill its halakhic obligations.

Notes

1. Jonathan Sacks, *One People? Tradition, Modernity, and Jewish Unity* (London: Littman Library of Jewish Civilisation, 1993), 27–31.
2. Tom Segev, *1949: The First Israelis*, trans. Arlen Neal Weinstein (New York: Free Press, 1986), 261.
3. Aviezer Ravitzky, "Is a Halakhic State Possible? The Paradox of Jewish Theocracy," *Israel Affairs* 11, no. 1 (January 2005): 137–164.

4. For example, Ira Sharkansky, *Rituals of Conflict: Religion, Politics, and Public Policy in Israel* (Boulder, CO: Lynne Rienner, 1996), 152.
5. Norman L. Zucker, *The Coming Crisis in Israel: Private Faith and Public Policy* (Cambridge, MA: MIT Press, 1973).
6. Charles S. Liebman and Eliezer Don-Yehiya, *Religion and Politics in Israel* (Bloomington: Indiana University Press, 1984), 51.
7. Ibid., 50.
8. Segev, *1949*, 208.
9. Ibid., 222.
10. Ibid.
11. Asher Arian, *The Second Republic: Politics in Israel* (Chatham, NJ: Chatham House, 1998), 27.
12. The Jews of Middle Eastern and North African origin are often called Sephardic as well as Oriental. Mizrachim has become the common term. For shift in Israeli identity, see Gershon Shafir and Yoav Peled, *Being Israeli: The Dynamics of Multiple Citizenship* (Cambridge, UK: Cambridge University Press, 2002).
13. Virginia Dominguez, *People as Subject, People as Object: Selfhood and People-hood in Contemporary Israel* (Madison: University of Wisconsin Press, 1989), 57.
14. Liebman and Don-Yehiya, *Religion and Politics*, 15.
15. Segev, *1949*, x.
16. Gidon Sapir, "Can an Orthodox Jew Participate in the Public Life of the State of Israel?" *Shofar: An Interdisciplinary Journal of Jewish Studies* 20, no. 2 (Winter 2002): 85.
17. Ibid., 238.
18. Rael Jean Isaac, *Israel Divided: Ideological Politics in the Jewish State* (Baltimore: Johns Hopkins University Press, 1976), 27.
19. Segev, 1949, 239.
20. Isaac, *Israel Divided*, 78.
21. Ibid., 78.
22. Segev, *1949*, 291.
23. Shafir and Peled, *Being Israeli*, 145, quoting Zucker, *The Coming Crisis*, 173.
24. Dominguez, *People as Subject*, 172.
25. Ibid., 172.
26. Shafir and Peled, *Being Israeli*, 318.

27. Elia Werczberger and Eliyahu Borukhov, "The Israel Land Authority: Relic or Necessity?" *Land Use Policy* 16, no. 2 (April 1999): 129–138.
28. Segev, *1949*, 258.
29. Ibid., 262.
30. Mizrachi is an abbreviation of "Mercaz Ruchani" or "spiritual center." See Bernard Avishai, *The Tragedy of Zionism: Revolution and Democracy in the Land of Israel* (New York: Farrar, Straus and Giroux, 1985), 95, for an account of its formation.
31. Shafir and Peled, *Being Israeli*, 138.
32. Segev, *1949*, 262.
33. Hanna Lerner, "Democracy, Constitutionalism, and Identity: The Anomaly of the Israeli Case," *Constellations* 11, no. 2 (June 2004): 239.
34. Ibid., citing Ralph Benyamin Neuberger, *The Constitution Debate in Israel* (Tel Aviv: Open University in Israel, 1997), 40.
35. Segev, *1949*, x.
36. Isaac, *Israel Divided*, 43.
37. Martin Edelman, "A Portion of Animosity: The Politics of the Disestablishment of Religion in Israel," *Israel Studies* 5, no. 1 (Spring 2000): 206. A 2007 law permits civil marriage for persons not considered Jews by Orthodox standards.
38. Menahem Parosh, quoted in Segev, *1949*, 261.
39. Aviezer Ravitzky, "Is a Halachic State Possible?"
40. Arian, *The Second Republic*, chapter 10.
41. Ibid., 3.
42. Yehudah Mirsky, "Inner Life of Religious Zionism," *New Leader* 78, no. 9 (December 4, 1995): 10–14.
43. Isaac, *Israel Divided*, 61.
44. Ibid., 62.
45. [Ralph] Benyamin Neuberger, *Religion and Democracy in Israel*, trans. Deborah Lemmer (Jerusalem: Floersheimer Institute for Policy Studies, 1997), 30.
46. Ehud Sprinzak, "Violence and Catastrophe in the Theology of Rabbi Meir Kahane: the Ideologization of Mimetic Desire," in *Terrorism and Political Violence* 3, no. 3 (Autumn 1991): 50.
47. Ibid., 54.

48. Raphael Cohen-Almagor, "Vigilant Jewish Fundamentalism: From the JDL to Kach (or 'Shalom Jews, Shalom Dogs')," *Terrorism and Political Violence* 4, no. 1 (Spring 1992): 54.
49. Ibid., 59.
50. Sprinzak, "Violence and Catastrophe," 58.
51. Cohen-Almagor, "Vigilant Jewish Fundamentalism," 51–52.
52. Efraim Ben Zadok, "State-Religion Relations in Israel: The Subtle Issue Underlying the Rabin Assassination," in Efraim Karsh, ed., *Israeli Politics and Society Since 1948: Problems of Collective Identity* (London: Frank Cass, 2002), 139.
53. Mirsky, "Inner Life of Religious Zionism."
54. Ben Zadok, "State-Religion Relations in Israel," 141.
55. Lisa Beyer, "The Religious Wars," *Time*, May 11, 1998, 32.
56. Alan Dowty, *The Jewish State: A Century Later* (Berkeley: University of California Press, 1998), chapter 2.
57. Lari Nyroos, "Religeopolitics: Dissident Geopolitics and the 'Fundamentalism' of Hamas and Kach," *Geopolitics* 6, no. 3 (Winter 2001): 135–157.
58. Sacks, *One People?* 61.
59. See, for example, Lerner, "Democracy, Constitutionalism, and Identity."
60. Edelman, "A Portion of Animosity," 207.
61. Ira Sharkansky, "Assessing Israel," *Shofar: An Interdisciplinary Journal of Jewish Studies* 18, no. 2 (Winter 2000): 5.
62. The state moved toward somewhat greater support in 2012.
63. Liebman and Don-Yehiya, *Religion and Politics*, 9.
64. Shafir and Peled, *Being Israeli*, 168.
65. Lilly Weissbrod, "Shas: An Ethnic Religious Party," *Israel Affairs* 9, no. 4 (Summer 2003): 96.
66. Etta Bick, "A Party in Decline: Shas in Israel's 2003 Elections," *Israel Affairs* 10, no. 4 (Summer 2004): 118.
67. Weissbrod, "Shas," 85.
68. Lerner, "Democracy, Constitutionalism, and Identity," 239.
69. Ibid., 244.
70. Ran Hirschl, "Constitutional Courts vs. Religious Fundamentalism: Three Middle Eastern Tales," *Texas Law Review* 82, no. 7 (June 2004): 1819–1870.
71. Lerner, "Democracy, Constitutionalism, and Identity," 247.

72. Frances Raday, "Women's Rights: Dichotomy Between Religion and Secularism in Israel," *Israel Affairs* 11, no. 1 (January 2005): 94.

73. Lerner, "Democracy, Constitutionalism, and Identity," 249.

74. Ibid.

75. Liebman and Don-Yehiya, *Religion and Politics*, 4.

76. Ibid., 7.

77. Ibid., 10.

78. Yeshayahu Leibowitz, *Judaism, Human Values, and the Jewish State*, ed. Eliezer Goldman (Cambridge, MA: Harvard University Press, 1992), 176.

79. Ibid., 227.

80. Joe Lockard, "Israeli Utopianism Today: Interview with Adi Ophir," *Tikkun* 19, no. 6 (November–December 2004): 21.

81. Shlomit Levy, Hanna Levinsohn, and Elihu Katz, "Believers, Observances and Social Interaction Among Israeli Jews," in Charles S. Liebman and Elihu Katz, eds., *The Jewishness of Israelis: Responses to the Guttman Report* (Albany: State University of New York Press, 1997), 3.

82. Ibid., 7.

83. Charles S. Liebman, "Academics and Other Intellectuals," in Liebman and Katz, eds., *The Jewishness of Israelis*, 69.

84. Bernard (Baruch) Susser, "Comments on the Guttman Report," in Liebman and Katz, eds., *The Jewishness of Israelis*, 170.

85. Gerald J. Blidstein, "The Guttman Report: The End of Commitment?" in Liebman and Katz, eds., *The Jewishness of Israelis*, 128; and Susser, "Comments on the Guttman Report," 169.

86. Levy, Levinsohn, and Katz, "Believers, Observances and Social Interaction," 21.

87. 87. Ibid., 31.

88. Menachem Friedman, "Comments on the Guttman Report," in Liebman and Katz, eds., *The Jewishness of Israelis*, 142.

89. Ronald Inglehart, *Modernization and Postmodernization: Cultural, Economic, and Political Change in 43 Societies* (Princeton, NJ: Princeton University Press, 1997).

90. Charles S. Liebman, "Cultural Conflict in Israeli Society," in Liebman and Katz, eds., *The Jewishness of Israelis*, 111.

91. Elihu Katz, "Behavioral and Phenomenological Jewishness," in Liebman and Katz, eds., *The Jewishness of Israelis*, 74.

92. Quoted in Liebman, "Academics and Other Intellectuals," Liebman and Katz, eds., *The Jewishness of Israelis*, 67.
93. Segev, *1949*, 202.
94. Julia Resnick, "Particularistic vs. Universalistic Content in the Israeli Educational System," *Curriculum Inquiry* 29, no. 4 (Winter 1999): 491.
95. Mordechai Bar-Lev, "Politicization and Depoliticization of Jewish Religious Education in Israel," *Religious Education* 86, no. 4 (Fall 1991): 609.
96. Resnick, "Particularistic vs. Universalistic Content," 492.
97. Zehavit Gross, "State-Religious Education in Israel: Between Religion and Modernity," *Prospects* 33, no. 2 (June 2003): 158.
98. Jo-Ann Harrison, "School Ceremonies for Yitzhak Rabin: Social Construction of Civil Religion in Israeli Schools," *Israel Studies* 6, no. 3 (Fall 2001): 113–134.
99. Martin Edelman, "A Portion of Animosity: The Politics of the Disestablishment of Religion in Israel," *Israel Studies* 5, no. 1 (Spring 2000): 215, put the number at 28,277, or 8 percent of all eligible draftees, in 1997.
100. The National Religious Party has founded some Hesder Yeshivot, a military-educational institution where students alternate between their studies and military service.

DISCUSSION QUESTIONS

1. What is promised in Israel's Declaration of Independence?
2. What did David Ben Gurion believe about theocracy and the Israeli State?
3. Explain the relationship between Orthodox Judaism and Labor Zionists?
4. What did the Law of Working Hours and Rest accomplish?
5. What events led to a militant religious Zionism ideology? Explain the idea of a Jewish settlement.
6. How did Yigal Amir justify the assassination of Prime Minister Yitzhak Rabin?

WOMEN

In Judaism, rabbis traditionally are male. Isaac Mayer Wise is considered the founder of American Reform Judaism, and although it has taken several decades this branch of Judaism has changed that all-male tradition. Reform Judaism now allows women to practice various rituals that, in more traditional branches of Judaism, are reserved for men. The first female rabbi was ordained in 1972. The following article discusses how issues like these developed and what connection they have to Hebrew Union College.

ARTICLE 3

A WORTHIER PLACE: WOMEN, REFORM JUDAISM, AND THE PRESIDENTS OF HEBREW UNION COLLEGE

By Karla Goldman

Every March the Hebrew Union College-Jewish Institute of Religion (HUC-JIR) observes Founders' Day, celebrating the founding presidents of the two institutes for rabbinical study that eventually merged to form the American Reform movement's present-day College-Institute. On Founders' Day the school's early leaders are remembered, and those rabbis who graduated from the College-Institute twenty-five years earlier are awarded honorary doctoral degrees in recognition of their years of service as rabbis. Until March 12, 1997, every rabbinical honoree at the Founders' Day celebrations held at the various branches of the College-Institute in Cincinnati, New York, and Los Angeles had been a man. But that particular March day was a special one, as it marked the first time at one of these occasions that a woman would be honored for having had the opportunity to serve the Jewish people as a rabbi.

Given the opportunity to speak on that Founders' Day in 1997, and aware that I was the first woman to speak at Cincinnati's annual commemoration, I could not help but reflect that Founders' Day is usually spent talking about men. There were certainly women in Cincinnati and elsewhere among those who supported the creation of the Union of American Hebrew Congregations in 1873 and Hebrew Union College in 1875. Hundreds of women from seventy-nine different localities, from Atchison, Kansas, to Zanesville, Ohio, joined Hebrew Ladies' Educational Aid Societies in 1877 by Paging to donate $1 a year "for the support of the indigent students of the Hebrew Union College."[1] Members of the National Federation of Temple Sisterhoods (now Women of Reform Judaism), founded in 1913, have supported Hebrew Union College since their organization's inception, providing many scholarships and funding the construction of a student dormitory that opened in 1925. As much as women may have contributed to framing the education of America's Reform rabbis, however, the hopes and direction that shaped the first one hundred years of Hebrew Union College belonged to men who first dreamed of and then realized the possibility of training young men to be rabbis for an American Israel. And so I, too, was compelled to turn to the vision of these male dreamers and builders.

Wise, Kohler, and Female Equality

Despite the clear male orientation of the founding and most of the history of Hebrew Union College, the question of women's religious equality and the possibility of female Jewish leadership threads an insistent pattern through the history of the school, raising challenges that still speak to the present moment's assumption of gender equality and apparent acceptance of female religious leadership. As it happens, Isaac Mayer Wise and Kaufmann Kohler, the first two presidents of Hebrew Union College, were among the earliest exponents of female equality within American Judaism. Early on in his American career Wise, who seemed to care about and comment on everything, identified the cause of women's rights as a useful weapon with which to push the cause of reform. In 1855 he argued for mixed choirs, suggesting that any objections to female participation could only "appear ridiculous" in light of contemporary sensibilities. "In our days," he argued, "only men who lacked full command over their intellectual faculties" could countenance the "queer notion of prohibiting ladies to sing in the choir." Further, Wise argued that it was "most desirable and recommendable that our ladies should take an active part in the synagogue." Such involvement could only add "to the devotion of the heart, the solemnity of the ceremonies, and the decorum of the divine place of worship."[2]

As the design of Cincinnati's Plum Street Temple, where he served as rabbi, suggests, Wise believed that a public and impressive Judaism should be expressed in grand sanctuaries that reflected the prosperous respectability of America's Jews. At the same time he expressed his disdain for the petty concerns of what he called "kitchen Judaism," denigrating the domestic religiosity that had so long been the domain of women's religious authority. It becomes more interesting in this context to see that although Wise may have eloquently articulated a call for women's equality within Judaism, he did not always implement his egalitarian suggestions.

Even though Wise was happy to take credit for pioneering the abolition of the women's gallery, family pews were not one of the reforms that he brought to Cincinnati when he arrived from Albany in 1854. It was not until Plum Street Temple opened in 1866, when family pews had already become part of the evolving national synagogue landscape, that mixed seating came to Cincinnati. Wise's inspiring call for a female college as a counterpart to his proposed men's seminary never found institutional expression, although his early college did host a number of girls and women in its courses.[3] In 1876 Wise called upon synagogue leaders to give women "a voice and a vote." Reform would not be "complete," Wise affirmed, until women were welcomed as members of congregations and congregational governing boards both "for the sake of the principle and to rouse in them an interest for congregational affairs."

Yet although he declared himself "ready to appear before any congregation" to plead the cause "of any woman wishing to become a member," his own congregation, over which he had a great deal of influence, never considered such a reform during his lifetime.[4] While Wise articulated a clear position of gender equality, he could not overturn the social, religious, and institutional boundaries that continued to limit opportunities for women within late-nineteenth-century American congregations.

Kaufmann Kohler, who became president of the college in 1903, had early in his career argued that, viewed historically, "only the Jewish religion assigned woman a worthier place" than other faiths by recognizing that her spiritual and "ethical nature" was equal to that of a man.[5] Kohler affirmed that Reform had a role to play in emancipating women from the oriental influences that had suppressed their public voices, and, like Wise, he insisted that "Reform Judaism will never reach its higher goal without having first accorded . . . equal voice to woman with man" in the congregational council and in the entire religious and moral sphere of life.[6]

At the same time, however, he revealed a profound discomfort with the possibility that bringing women's voices into public Jewish life might add to the dimming of "religion's fire" that "has almost burned out on the domestic altar."[7] Although the Reform project that Kohler championed was driven by the force of reason, at the same time he yearned for the "beautiful Sabbath lamp of household piety and devotion," a Judaism that touched the soul and heart.[8] The problem with modern society, Kohler argued, was that "the blood has all rushed up to the brain and the heart emptied of its vital fluid has its chills and fevers." Kohler looked to women for a solution but not to "the intellectual woman," whose circulation also sent "blood rushing to the head and away from the heart." Rather, it would be "the hearty, the whole-souled tender-hearted Rebecca-like woman" who would finally bring "the lost paradise back to man."[9]

Kohler rejected the idea of a woman who might not fit his definition. For him the threat of meaninglessness in a posttraditional world was exemplified in the figure of a woman refusing to play her proper role. Even the possibility of such a creature seemed profoundly disturbing to Kohler: "A woman without tenderness, without gentleness, without the power of self-suppression to an almost infinite degree is a creature so anomalous that she cannot fail to do enormous harm, both to her own sex and the other. . . . She . . . becomes a devil in disguise."[10]

When Kohler looked to women to maintain familiar, if updated, feminine roles, he imposed upon them the responsibility for maintaining the distinctiveness of Judaism. In his question "Is it woman's calling to become a man?" Kohler struggled with the need to place limits upon certain essential markers of identity.[11] Although Reform Judaism sought to meet the modern needs of both men and women, the only vital future Kaufmann Kohler could envision for Judaism was one in which women would continue to embody and sustain the Jewish past.

Pushing the Reform Establishment

For all their engagement in questions of women's religious status, neither Isaac M. Wise nor Kaufmann Kohler had to seriously consider the question of women's religious leadership. This was not the case for their successors. Soon after Kohler's retirement, Martha Neumark's pursuit of studies at Hebrew Union College and her desire to serve a High Holy Days student pulpit pushed the Reform establishment to confront practical questions regarding the proper callings for men and women within Judaism. In an extended discussion by the Central Conference of American Rabbis (CCAR) in 1922 Kohler's successor, Julian Morgenstern, excused himself from "express[ing] any opinion upon this subject," although he did encourage the busy conferees to expend "as much time as may be necessary for a thorough discussion of the question." Morgenstern asked them to look at the question pragmatically: "Namely, is it expedient, and is it worth while?"[12] In the end the 1922 conferees did resolve that given Reform Judaism's other departures from tradition, women could not "justly be denied the privilege of ordination." HUC's board of governors, however, chose to overturn this less than emphatic affirmation, arguing that the question was not one of practical import.[13]

At New York's Jewish Institute of Religion Stephen Wise enthusiastically acceded to the request of Irma Levy Lindheim to enter the course of study at his new institution in 1922. The faculty's response to Lindheim's subsequent petition to change her status from that of special student to that of a regular student was reflected in the school's surprising 1923 charter, which, in describing the institute's mission to "train, in liberal spirit, men and women for the Jewish ministry, research, and community service," perhaps suggested the faculty's view that women students should be training themselves for something other than the rabbinate. In any case Lindheim withdrew from the curriculum before the issue of her ordination could become a reality, and when Helen Levinthal completed the JIR curriculum in 1939, faculty reluctance to certify her as a rabbi resulted in her being granted a master of Hebrew letters degree rather than ordination.[14]

A 1956 report to the CCAR submitted by a committee that included HUC-JIR president Nelson Glueck stated clearly that the conference should "endorse the admission" of women students to the College-Institute. This suggestion was laid aside, however, purportedly until the other side of the issue could be considered.[15]

The Reform movement's hesitation in following through on its stated commitment to the realization of women's equality might have continued indefinitely. But as societal expectations of women's roles shifted in the 1960s, and as Sally Priesand advanced through the Cincinnati campus's undergraduate and rabbinic course, the question was finally resolved without any grand statements from the movement's governing bodies. Nelson Glueck made it clear that he would ordain a female candidate for ordination when the opportunity arose. In 1972, after Glueck's death, Alfred Gottschalk was able to carry through on his predecessor's commitment.

A Generation of Women Rabbis

Now, at the start of a new century, we can literally say that we have had women rabbis or at least a woman rabbi for a generation. Clearly HUC-JIR's step in allowing women to take on the most prominent role of Jewish religious leadership has had profound symbolic and practical implications for the Jewish people, with an impact that resounds far beyond the rabbinate and the Reform movement. The Reform movement's innovation was followed quickly by the Reconstructionist Rabbinical College and, eleven years later, by the Jewish Theological Seminary. The presence of women in Jewish leadership positions has reconfigured expectations of what women should be allowed and encouraged to do in a wide variety of Jewish settings. As women have responded to the range of possibilities so recently opened to them, a worthier place for women has been created in the public realm of Judaism.

In keeping with this achievement and cognizant of remaining challenges, the Statement of Principles for Reform Judaism, adopted by members of the Central Conference of American Rabbis when they met in Pittsburgh in 1999, pledged "to fulfill Reform Judaism's historic commitment to the complete equality of women and men in Jewish life." The spirit of this commitment would have been familiar to Wise, Kohler, Morgenstern, and Glueck. Indeed, although the leaders of today's Reform movement face an order of challenges different from those encountered by their predecessors, they must also struggle to bring reality into conformity with their rhetoric.

We are only beginning to recognize the implications of what happens when, to paraphrase Kohler, a man's calling also becomes a woman's. It is no great secret that despite the fact that the College-Institute has been ordaining women at a rate that for the last fifteen years has approached parity with that of men, we have not put to rest the problem of women's exclusion within our tradition. In fact, at times it seems that all these many women rabbis have done is stir up a lot of new questions.

Once women gained the right of Jewish religious leadership, some among them, almost inevitably, began to question the male orientation of the prayers they were now empowered to lead and the traditions they were now called upon to interpret. Although the Reform movement placed itself in the forefront of realizing and embracing the challenges of a truly progressive Judaism when it ordained Sally Priesand, since that day many of the more radical attempts by women to claim their place within Jewish tradition and practice have arisen outside the Reform movement. More than twenty-five years into this change, some feminist liturgical pioneers have introduced feminized pronouns and verb forms to refer to both God and the worshipers in the Hebrew liturgical text; others seem to have removed God entirely. Feminist scholars have focused a critical eye on a profoundly patriarchal and hierarchal tradition that, they suggest, will not suddenly become acceptable if we simply omit or modify teachings and practices that explicitly exclude or denigrate women. Gay activists both within and beyond the Reform movement have built upon feminist claims and achievements, questioning whether the Jewish marriage sacrament of kiddushin should be offered only to heterosexual couples. They want to know why gays and lesbians should be made to feel less-than-equal claimants to the richness of Jewish life.

These challenges have found a mainstream response from within the Reform movement with the creation of gender-sensitive liturgies that insert references to Sarah, Rebecca, Leah, and Rachel into prayers that refer to the patriarchs and, in the English text, carefully avoid projecting male identities upon either God or humanity. Future liturgy projects promise even more compelling engagement with the challenges raised by feminist liturgical work. Movement policy statements, meanwhile, have accepted gay rabbis (while carefully specifying that only heterosexual choices should be understood as constituting a Jewish ideal) and have called for the civil legitimization of lifelong gay marriages or partnerships. In March 2000 the CCAR further challenged traditional Jewish definitions of sexuality and family by expressing its affirmation both of the decisions of rabbis who choose to officiate at "rituals of union for same-gender couples" and the decisions of those who choose not to officiate.

In addition to forcing us to reevaluate our relationship to traditions that have been created by generations of male authorities, the presence of women religious leaders has also challenged us to reconsider a professional culture that was built around the lives and life course of men. Reform leaders at one time may have wished to believe that all that was necessary to create women rabbis was to allow women to complete the course of study (at a school that for one hundred years had been designed to train men), to try to teach everybody fairly, and to hire, eventually, a few women professors.

All of this has now been achieved, but before we rest satisfied, we must ask a few difficult questions. Is the model of a successful rabbinate that prevailed when only men were rabbis still accepted as the appropriate model for success? Do HUC-JIR rabbinical students have the opportunity to learn from teachers who will challenge the male orientation of our traditional texts? Are students and congregants within the Reform movement given the opportunity to learn from women's voices? Do all within the umbrella of the movement's various institutions feel that their voices are heard, that their struggles are recognized, and that they are part of a movement that can respond to their needs? Can we recognize that we may have something to learn from voices that may not sound like our own? Are the questions of those who challenge prevailing patterns heard and treated with respect? Are challenges to our liturgies, our texts, and our theologies—challenges that grow out of a desire to understand how Jewish tradition has belonged to women as well as men—treated seriously as important questions for both men and women? We cannot yet consistently answer these questions affirmatively.[16]

All this questioning, for all of the creative energy that comes along with it, may indeed prove too challenging. When Kaufmann Kohler lost the grounding of the absolute faith of his youth, he cast about for something to hold on to that could continue to tell him who he was. Trying to live in a globalizing economy and amid dizzying technological and social change, even those of us who are committed to a progressive and liberal Judaism may still hope that religious faith can offer simple constructs of meaning and guidance. We, too, would like assurances of where we stand and who we are. In this context feminist challenges to text, tradition, ritual, and theology seem to undermine all that was familiar, everything that told us who we were as Jews. Suddenly we sympathize with Kohler! Can we truly challenge boundaries and hang on to meaning? Can we blur certainties yet remain committed to Judaism? There is nothing simple about this. We can understand why Kohler wanted to hang on to his certainties, why Wise would not push for reforms that might have represented too much of a challenge to the constituencies that he served. When there is already so much work to do, it can seem easier to reject the burden of additional complexity.

When the time came to honor a generation of service from those who became rabbis in 1972, I tried to imagine myself at their ordination service at Plum Street Temple. I wondered what it must have felt like to know that a moment of profound personal and individual import was also a moment that redefined the parameters of Jewish life. Although they sat in his sanctuary, Sally Priesand and her classmates were hardly contemporaries of Isaac Mayer Wise. Yet as they symbolically accepted the burden of changing the world for all of us, they, too, became founders—they helped shape and were part of a moment that redefined the history not only of the American Reform movement, but of Judaism itself.

To the men and women who study for the rabbinate today, Sally Priesand's journey to ordination began a world away in time and experience. What a difference for a girl who grows up wanting to be a rabbi to know that she can be one. But we do a dishonor to the legacy of Sally Priesand and those who supported her if we sit back and assume that these founders did all that needed to be done. If they cannot complete this work, neither are any of those who profess commitment to a progressive Judaism free to desist from it.

Once women's presence was legitimated in the halls of Hebrew Union College-Jewish Institute of Religion, once women rabbinical students and women rabbis began to redefine notions of Jewish leadership, they found a worthier place for themselves within Jewish tradition than Kaufmann Kohler could ever have imagined. At the beginning of the twenty-first century, with the highly symbolic barrier of women's access to formal spiritual leadership irrevocably breached, today's Reform leaders, students, and congregants must grapple with the evolving challenges that continue to arise from that transformative moment at Plum Street Temple.

If they refuse to believe that the work is done or to accept complacency, if they continue to challenge the assumptions of both traditional and Reform Judaism, if they inherit the struggle pioneered by those rabbis ordained in 1972, then they will do more than merely create a worthier place for those in their own generation. They will follow the course of earlier leaders, and they will emulate the continuing work of Sally Priesand and her classmates. They will struggle to create an American Reform Judaism that can provide a worthier home, a worthier place for those who long for a Judaism that can honor not only our past, but also our future.

Notes

1. Ladies' Educational Aid Societies, "Roles of Honor," Box X-206, American Jewish Archives, Cincinnati, Ohio.
2. Isaac M. Wise, "Does the Canon Law Permit Ladies to Sing in the Synagogue?" *Israelite* 2, 5 (August 10, 1855) and 6 (August 17, 1855): 44–45.
3. Isaac M. Wise, "What Should Be Done?" *Israelite* 1, 4 (August 4, 1854): 29. For a detailed portrayal of the different generations of female students at Hebrew Union College and the Jewish Institute of Religion who preceded Sally Priesand, see Pamela S. Nadell, *Women Who Would Be Rabbis: A History of Women's Ordination*, 1889–3985 (Boston: Beacon Press, 1998).
4. Isaac M. Wise, "Woman in the Synagogue," *American Israelite,* September 8, 1878.
5. Kaufmann Kohler, "Das Frauenherz oder das Miriambrunnlein im Lager Israels," *Jewish Times* 2, 5 (February 17, 1871): 812. On Kohler's ambivalent stance on the role of women in

modern Judaism, see my article "The Ambivalence of Reform Judaism: Kaufmann Kohler and the Ideal Jewish Woman," *American Jewish History* 79 (Summer, 1990): 477–99.

6. Kaufmann Kohler, conference paper, in "Proceedings of the Pittsburgh Rabbinical Conference, 1885," in *The Changing World of Reform Judaism: The Pittsburgh Platform in Retrospect*, ed. Walter Jacob (Pittsburgh: Rodef Shalom Congregation, 1985), 96.

7. Ibid., 96, 102.

8. Kaufmann Kohler, "Are Sunday Lectures a Treason to Judaism," *Temple Beth–El Sunday Lectures* 16 (January 8, 1888): 7–8.

9. Kaufmann Kohler, "Rocks Ahead," *Reform Advocate 2,* 13 (November 14, 1891): 215; Kaufmann Kohler, "The Jewish Ideal of Womanhood," 1899, Box 6 / Folder 2, Kaufmann Kohler Papers, Coll. 29, American Jewish Archives, Cincinnati, Ohio.

10. Kaufmann Kohler, "Esther or the Jewish Woman," *Temple Beth–El Sunday Lectures* 16 (February 26, 1888): 2.

11. Kaufmann Kohler, "Der Beruf des Weibes," *Jewish Times,* May 21, 1875, J88.

12. Central Conference of American Rabbis (CCAR) *Yearbook* 32 (1922): 167.

13. See Ellen M. Umansky, "Women's Journey toward Rabbinic Ordination," in *Women Rabbis: Exploration and Celebration, Papers Delivered at an Academic Conference Honoring Twenty Years of Women in the Rabbinate,* 1972–2992, ed. Gary P. Zola (Cincinnati: HUC-JIR Rabbinic Alumni Association Press, 1996), 32–33.

14. On Lindheim, see Pamela Nadell, "The Women Who Would be Rabbis," in *Gender and Judaism: The Transformation of Tradition,* ed. T. M. Rudavsky (New York: New York University Press, 1995), 127–31 and Nadell, *Women,* 72–76; on Levinthal, see Nadell, *Women* 80–85, and Umansky, *Women's Journey,* nn. 22, 40.

15. CCAR *Yearbook* 68 (1956): 93.

16. For other reflections on the difficulty of realizing the implications implicit in the creation of women rabbis, see Shulamit S. Magnus, "The New Reality: The Implications of Women in Rabbinical School," *Reconstructionist,* summer 1991, 16–18; Beth Wenger, "The Politics of Women's Ordination: Jewish Law, Institutional Power, and the Debate over Women in the Rabbinate," in *Tradition Renewed: A History of the Jewish Theological Seminary,* ed. Jack Wertheimer (New York: Jewish Theological Seminary, 1997), 2: 483–523. Important observations are also contained in many of the essays included in Zola, ed., *Women Rabbis.*

DISCUSSION QUESTIONS

1. How did Isaac Mayer Wise fight for the rights of Jewish women? What limited his attempts?
2. How did some of the earlier presidents of Hebrew Union College view the ordination of women?
3. Explain the process of achieving ordination for women in Reform Judaism.
4. What did the Central Conference of American Rabbis declare in 2000 concerning same-sex marriage?
5. How was the anecdote about Plum Street Temple a transformative moment in Reform Judaism?
6. Who is Sally Priesand?

CHRISTIANITY

INTRODUCTION

What today is called Christianity was at first another form of Judaism. This early Jewish group believed that the foretold coming of the Messiah, cited in several prophetic writings in the Hebrew Bible, was fulfilled in **Jesus Christ**. In the Hebrew Bible, **Messiah** refers to "a man who will restore the customs, practices, and worship of God for the children of Israel." In the first century BCE, the children of Israel, held under the iron fist of Rome, hoped greatly for the coming of this man. Early Christian literature demonstrates that some of these Jewish people believed the end of the world was imminent. However, the title of "Messiah" comes with different roles inside of what would become Christianity, and because of it, many Jewish people believe Jesus of Nazareth falsely claimed to be the Messiah. Many forms of Judaism are still waiting in the coming of the Messiah, but for most Christians today, Jesus is not only the Messiah but also the incarnated form of God.

Although Christianity is considered a monotheistic religion, many forms of Christianity understand this concept as working through a Trinity. The Trinity was proclaimed in some early forms of Christianity, as can be seen in the **Apostles' Creed** and the **Nicene Creed**. The **Trinity**, a paradox for many believers, holds the idea that "one God exists in three forms: Father, Son, and Holy Spirit." The relationships among the three are a difficult topic to broach. A quick, generalized explanation is that the spirit of God, **the Father**, came down to Earth and incarnated within the human form of Jesus, **the Son**. Many Christians then believe that Jesus died on a cross—with his death caused not by the Roman powers that crucified him but by his own will, in order to atone for the sins of all humankind. His sacrifice became a bridge across which people who believe can access the presence of God within themselves—that is, through **the Holy Spirit**. The idea is that the spirit of God dwells in different places, first within Jesus and then across the billions of people who have become believers since then.

The early history of Christianity was tumultuous, and it changed drastically within only a few centuries. Christianity went from being a part of Judaism to becoming its own universal religion, and from enduring heavy persecution by the Roman Empire to being the official religion of that same empire. **Paul** became a key figure in the process of transitioning away from Judaism and then in the eventual growth of what is now the Christian Church. In fact, Paul is known as the seconder founder of Christianity. Many of his letters are considered to be some of the earliest Christian literature, and they have been canonized in what is known as the **New Testament**. In these letters, which make up about half of the canon, he developed his ideas about this new community and also provided an in-depth theological analysis of who Jesus was. In Paul's theology it was God's plan that the **gentiles**, or "people who were not Jewish," would join the fold. Further, much of the Jewish ritual practice would fade out and in some instances be replaced with rituals specifically remembering and honoring Jesus.

Paul was active in traveling his known world to spread the ideas and words of Jesus and start new churches. Many of his travels, documented in Christian literature, were met with great hardship, including shipwrecks, stoning, and imprisonment—and through it all, an unwavering faith. The earliest beliefs and communities within what would come to be known as Christianity are not uniform, however. There were many different communities, and they processed who Jesus Christ was in many different ways. Most likely, there was a majority and several minorities. When **Constantine** became emperor of the Roman State in the fourth century CE, he began to change the lives of the Christians, many of whom were under intense persecution by the Roman emperor **Diocletian** at the beginning of that century. Constantine passed edicts that called for tolerance toward the Christians, and he was the first Roman emperor to convert to the Christian faith.

Much as **Emperor Ashoka** did within Buddhism centuries earlier, Constantine convened councils to attempt unification within the distinct early Christian communities. The councils decided many issues, including the nature of Jesus Christ, and they drew up an early list of standardized and normative texts for this faith. Whether his attempts were for unity or for solidification of power, Constantine paved the way for Nicene Christianity to become the official state religion of the Roman Empire under **Theodosius I**. Subsequently, the history of this faith took many turns. At one point, it was governed by bishops in five prominent cities, in an arrangement known as the **pentarchy**. From one of these locations, the city of Rome, the institution grew to become the **Roman Catholic Church**. Some of the cities and churches were included in this church, led by the Pope; others were not. In 1054, the tension in leadership led the Roman Catholic Church to officially separate from the **Eastern Orthodox Church**, in an event known as the Great Schism.

About 500 years later, a European movement known as the **Protestant Reformation** rose up against the Roman Catholic Church and demanded reforms in accessibility, control, and power. The result was another divide—and what became the **Protestant Church**. This church exists as its own branch of Christianity, with countless different **denominations**, the number of which is still growing annually. Each denomination has unique understandings about the Bible and what it means for application and practice. Some of the earliest Protestant denominations translated the Bible into languages other than Latin, so that people could read the Bible for themselves. Further, the Protestant denominations developed different clerical systems that followed different practices in their leadership and rituals. These systems are followed today, with some denominations denying the Trinity, others refusing to fight in wars, and others believing that healing comes from prayer alone.

Many Protestant denominations argue over correct doctrine and theology, and they have developed specific forms of baptism and communion. These two rituals are essential to the practice of most forms of Christianity, although they may have originated in Judaism. For example, baptism may have emerged from the Jewish ritual of ***mikvah***, or "water immersion," and communion may have come from the ritual dinner of the Passover *seder*. In most forms of Christianity, baptism and communion are public acts of obedience and faith. Both are powerful rituals meant to connect the believer with Jesus. There is supposed to be a deep level of communing and intimacy with Jesus. For many believers, Jesus is the central figure and the bridge to the divine; ideally, a Christian will reach full identification with Jesus. The goal for many is to think and act as Jesus did.

Baptism reenacts the sacred story of Jesus's own baptism. It also creates a sense of transformation in the believer, as there is ritual purification through the water. For many Christians, especially in newer Protestant denominations, baptism brings a feeling of complete identification with Jesus: They are now crucified with him. They believe they crucify the person they once were, in order to resurrect with Jesus and be born again; they believe they have new life in Jesus. **Communion**, like most rituals, also reenacts a sacred story. During a Passover *seder*, Jesus asked his followers to partake of bread and wine. The bread and wine represent the body and blood of Jesus, who would be broken and crucified. In the first century BCE, eating was an intimate act, and today it still serves as an intimate form of communing with Jesus. The person "taking communion" is literally taking in what most Christians see as a sacrifice that means the love of God, eternal salvation, and life.

ECOLOGY

Across Christianity, there are different understandings of environmentalism. Historically, some older Protestant denominations have even been connected to early environmentalist movements. However, many people view Christian theology as having a negative impact on ecological concerns. The following article explores the problematic theologies that might propel this dynamic.

ARTICLE 1
FUNDAMENTALIST DOMINION, POSTMODERN ECOLOGY

By Paul Maltby

Christian fundamentalist hostility to environmentalism typically finds its endorsement in the book of Genesis. A literal reading of the injunction that "man" should "replenish the earth, and subdue it: and have dominion over the fish of the sea, and over the fowl of the air, and over every living thing that moveth upon the earth" (1:28) has ratified the view of nature as a God-given resource for unlimited human use. This view was provocatively expressed by Ann Coulter, the right-wing Christian radio talk-show host, when she observed, "God gave us the earth. We have dominion over the plants, the animals, the trees. God said, 'Earth is yours. Take it. Rape it. It's yours'" (*Hannity*). Evidently, dominionist philosophy does not recognize natural entities and species as autonomous life-forms; rather, it perceives them as artifacts designed to satisfy human needs. Indeed, according to the fundamentalist economist E. Calvin Beisner, to put the earth before human needs is to be guilty of "idolatry of nature" (*Prospects* 165). (Without citing Saint Paul, Beisner surely has in mind the Epistle to the Romans, in which Paul condemns the ungodly and wicked who "worshipped and served the creature rather than the Creator" [1:25].)

More than forty years ago, Lynn White Jr. examined the ecological implications of the "Genesis creation story" in his famous essay "The Historical Roots of Our Ecologic Crisis." He argued that the key premises Christians derived from Genesis are (1) God planned the Creation for the benefit and rule of humanity; (2) insofar as man is made in God's image, he exists outside of nature. Thus, given "the Christian dogma of man's transcendence of, and rightful mastery over, nature" (90), the humanity/nature dualism was firmly established. Moreover, compared with pagan animism and Eastern religions, White insists, "Christianity is the most anthropocentric religion the world has seen" (86). Such is the thinking that has facilitated the super-exploitation of the natural world, for which "Christianity bears a huge burden of guilt" (90). However, White does not overlook Saint Francis of Assisi, whom he extols as "the greatest spiritual revolutionary in Western History" (93), on the grounds that the saint's "belief in the virtue of humility . . . for man as a species" led to his effort to "depose man from his monarchy over creation and set up a democracy of all God's creatures" (91). But this doctrine was suppressed by the Church, and White concludes: "We shall continue to have a worsening ecologic crisis until we reject the Christian axiom that nature has no reason for existence save to serve man" (93). In short, in his classic argument, White anticipated several themes central to conventional ecological or postmodern critiques of Christian dominionist attitudes to nature: namely, the humanity/nature dualism, anthropocentrism, the virtue of humility in relation to nature, and the democracy of species.

Broadly speaking, dominionism means the responsibility of Christians to subject the spheres of everyday life and all institutions to the rule of God's laws, thus (from the postmillennialist standpoint) securing the conditions for Christ's return. In environmentalist debates, however, the use of the term is limited to designate the belief that the achievement of a sovereign and exploitative power over nature is mandated by the Bible. Still, the distinction is somewhat tenuous. After all, for the powerful Reconstructionist wing of Christian fundamentalism, the broad definition of dominionism is derived from the limited definition. As Sara Diamond explains in *Spiritual Warfare,* Reconstructionists such as the late R. J. Rushdoony and Gary North interpret dominionism as meaning "Christians are Biblically mandated to 'occupy' all secular institutions" until Christ returns (138). Dominion theology is about "subduing the earth" in the name of Christ, a subjugation that extends from nature, via a stringent application of biblical laws, to all forms of sociocultural existence. In short, dominion is a divine right, lost by Adam with the Fall but reclaimable by born-again Christians.

Fundamentalist antienvironmentalism finds further support in a literal reading of New Testament accounts of the Apocalypse and the Rapture as prophecy (Matt. 24:7, Luke 21:8, Rev. 8:8–11, 1 Thess. 4:16–17). Here, interpretation demotes nature to a mere prop in the supernatural drama of human salvation. Thus, not only is conservation seen as irrelevant, insofar as the planet is thought to have no future (in the words of the nineteenth-century premillennialist Dwight Moody, "You don't polish the brass of a sinking ship"), but environmental catastrophe is positively welcomed by Pat Robertson and other fundamentalist leaders as presaging the Rapture and the Second Coming. This is the End Times doctrine that permeates the books of Tim LaHaye and Jerry Jenkins and of Hal Lindsey, where ecological disaster is crucial to the plot of the Left Behind series (e.g., LaHaye and Jenkins *Left Behind,* 311–12) and to the good news in *The Late Great Planet Earth* (Lindsey 166). The same doctrine has motivated the congressional antienvironmentalism of leading Republicans: Tom DeLay, House majority leader from 2003 to 2006; James Inhofe, currently the senator for Oklahoma and, given President George W. Bush's antienvironmentalism, chairman of the Senate Committee on Environment and Public Works from 2003 to 2007; and James Watt, who served as Ronald Reagan's secretary of the interior from

1981 to 1983. Todd Strandberg, webmaster of RaptureReady.com, the website that relays news of environmental catastrophes in the context of biblical prophecies of the Apocalypse, has observed that "global warming could very well be a major factor in the plagues of the Tribulation" (qtd. in Scherer). The guiding thought for all these figures is, why care about ecological crisis when true believers will be rescued by the Rapture?

Christian Reconstructionist politics also drives the zealous antienvironmentalism of the fundamentalist leadership. Here, criticism of environmentalism generally adopts two strategies: either it misrepresents the environmentalist movement as dominated by radical leftists and eco-pagans, or it scours the margins of the environmentalist debates from which to enlist "experts" who challenge the claims of the mainstream scientific community. Thus, Pat Robertson has identified those calling for the "empowerment of ecology" as advocates of a "one-world socialist government" (*New World* 153, 215). (The fundamentalists' political suspicion of environmentalists has prompted them to brand the latter as "watermelons," i.e., green on the outside, red within.) For Richard T. Wright, a professor of biology at Gordon College, Massachusetts, "Christian anti-environmentalism can be traced directly to political commitments." He considers the arguments that question the soundness of environmental science or that see a left-wing conspiracy behind environmentalism to be "red herrings." He concludes, "The political right has lost its traditional enemy—world communism—and appears to be replacing it with world environmentalism. The Christian political right is following right along the party line" (89, 90).

The most outspoken critics of environmentalism from the fundamentalist fold are E. Calvin Beisner and the late Larry Burkett. Burkett, a popular author of books that promote Christian principles of financial management, concentrated much of his fire on environmental regulations. Like Robertson, he saw environmentalists as communists in disguise, whose aim is to enlarge central government by exaggerating environmental problems. Government-authorized environmental programs operate "in the same role as the KGB" and threaten the vitality of American business: "The EPA [has] become a paramilitary enforcement group running amuck throughout the free enterprise system" (178–79). Moreover, as he saw it, environmental regulations constitute a huge burden for the economy. He believed that "environmental extremists" have overrun the political system and are implementing policies at odds with the Christian reconstruction of society. However, Burkett did not rely on the Bible to endorse his position but on dubious research from non-refereed "scientific" sources, such as articles published in the *New American,* a journal sponsored by the John Birch Society.

E. Calvin Beisner, a professor of interdisciplinary studies at Covenant College, Georgia, and adjunct scholar at the Acton Institute, is a key signatory to the (deceptively named) "Cornwall Declaration on Environmental Stewardship" (ICES). Like Burkett, he affirms his opposition to the scientific consensus on the causes of environmental crises with appeals to the work of maverick climatologists and other marginal researchers. However, he also enlists support for his antienvironmentalist stance by adducing passages from the Bible. Citing the dominion mandate in Genesis, he avers, "Man was not made for the earth, but the earth for man" (Prospects 24, 163). Invoking Psalms 115:16 and 8:6, he sees "man as [God's] vice-regent" (156), a subordinate owner of the earth rather than its steward. He argues, "God intended there to be considerable liberty regarding the ways in which we rule the earth" (163), and maintains that the unregulated and private use of resources is consonant with God's provision of a bountiful earth to serve humanity's needs. Furthermore, he insists that "global warming [is] indeed an expression of God's will" (qtd. in Moyers); that is to say, the destructive effects of climate change are not primarily the result of irresponsible social practices but God's punishment for human sin, comparable to the flood in the story of Noah. (God cursed the earth after Adam and Eve, so natural disasters like that of Hurricane Katrina are expressive of God's punitive will.) Another of Beisner's responses to environmental devastation is to quote Saint Paul in 1 Corinthians: "For I determined not to know anything among you, save Jesus Christ, and him crucified" (2:2), which he interprets as meaning that human experience on earth is insignificant compared to the eternity of our salvation in Christ. He dismisses concerns about the depletion of natural resources, arguing that (1) God's bounty is infinite and (2) God-given human creativity will enable us to solve the problem of resource scarcity.

Fundamentalists vigorously contest ecological concerns about resource depletion. Mark Beliles and Stephen McDowell, authors of the fundamentalist high school textbook *America's Providential History* (1989), are in no doubt about the bounty and dependability of God's providence: "The secular or socialist has a limited resource mentality and views the world as a pie (there is only so much) that needs to be cut up so that everyone can get a piece. In contrast, the Christian knows that the potential in God is unlimited and that there is no shortage of resources in God's Earth. The resources are waiting to be tapped" (197). Here, one must observe, such an argument, squarely based on the fundamentalist faith in "America's providential history," surely finds reinforcement from a residual colonial mentality: that lingering perception of the New World as a space of infinite resources available for plunder.

However, it is important to stress that evangelicals can be found in both the environmentalist and antienvironmentalist camps. Organizations such as Evangelicals for Social Action and evangelical journals such as *World Vision* and *Moody Monthly* proceed from alternative passages in the Bible to actively promote an ethic of environmental stewardship. They most frequently cite Genesis 2:15: "And the LORD God took the man, and put him in the garden of Eden to dress it and to keep it." According to the Reverend Richard Cizik, vice president for governmental affairs of the 30-million-strong National Association of Evangelicals, 63% of evangelicals acknowledge the reality of climate change. And though a self-described "pro-Bush conservative," Cizik lobbies hard on behalf of environmentalist causes, integrating environmentalism into the NAE's political agenda. He invokes Genesis 2:15 and Revelation 11:18 (God will "destroy them which destroy the earth") in support of his pro-environmentalist stance.

Cizik's efforts are reinforced by the Reverend Jim Ball, executive director of the Evangelical Environmental Network. The EEN's flagship publication, *Creation Care,* provides "biblically informed and timely articles" on environmental issues, in particular detailed essays and fact sheets on pollution and global warming. In 1994 the magazine published the highly influential "An Evangelical Declaration on the Care of Creation" (Adeney et al.). Bluntly acknowledging many forms of environmental damage, the manifesto declared the Christian obligation to care for the Creation and argued that biblical faith is essential to the solution of ecological problems.

Cizik and Ball are among the most prominent voices speaking for grassroots evangelical communities that have challenged dominionism in the name of "Creation care" (a term which calls for responsible stewardship of the earth but which, unlike "environmentalism," is free of secular left-liberal overtones). Yet, for all their lobbying efforts, it is the fundamentalist top brass, most of whom belong to the extreme Christian Right—notably, Pat Robertson, Ralph Reed, and, above all, Cizik's bête noire James Dobson[1]—who, at least during Bush's two terms, held more sway over the government's environmental legislation. They enjoyed privileged access to White House lawmakers, including and especially Gale Norton, Bush's secretary of the interior from 2001 to 2006 (Kaplan 81), who in the 1980s worked under James Watt and is a vocal advocate of free-market environmentalism.

In April 2000, provocatively timed to coincide with the thirtieth anniversary of Earth Day, a coalition of fundamentalist and theologically conservative religious groups launched ICES, the Interfaith Council of Environmental Stewardship. Bill Berkowitz, editor of *Culture Watch*, has described ICES as "an organization to graft dominion theology onto right-wing environmentalism." ICES's founding document, the "Cornwall Declaration on Environmental Stewardship," expounds ideas largely derived from Beisner's *Where Garden Meets Wilderness* (1997). This document significantly understates the global scale of environmental crises and seeks to characterize humans as divinely mandated "producers" instead of reckless "consumers and polluters." The author of the document elaborates thus: "Our call to fruitfulness . . . is not contrary to but mutually complementary with our call to steward God's gift s. This call implies a serious commitment to fostering the intellectual, moral, and religious habits and practices needed for free economies and genuine care for the environment." Clearly, in this context, the eco-friendly concepts of "stewarding" and "genuine care for the environment" and the innocuous-sounding "call to fruitfulness" simply serve to make palatable the ideological commitment to unregulated exploitation of the planet's resources. Indeed, the declaration concludes: "We aspire to a world in which widespread economic freedom—which is integral to private, market economies—makes sound ecological stewardship available to ever greater numbers" (ICES). Here, the aspiration is purely rhetorical insofar as "sound ecological stewardship" cannot be privately enforced; time and again, private enterprise altogether avoids or simply reneges on "voluntary" commitments to respect the environment.[2] In short, the ICES document engages in the tactics of greenwashing, that is, terminology that enables antienvironmentalist policies to masquerade as environmentally friendly. (Cf. the Bush administration's "Healthy Forests Initiative" [USDA] and "Clear Skies Initiative" [EPA]: eco-friendly designations that mask policies of deregulation, allowing for, respectively, more deforestation and pollution.)

ICES is the inspiration and brainchild of the Michigan-based Acton Institute for the Study of Religion and Liberty. The institute was founded by Father Robert Sirico, in 1990, as an organization that advocates the use of the Bible in conjunction with free-market economics as a guide to policy making. So it should come as no surprise to learn that, in May 2003, Sirico was present at Exxon-Mobil's annual shareholders' meeting, where he spoke out against environmental resolutions proposed by religious social activists (Berkowitz). Moreover, for years the Acton Institute has received funding from Exxon-Mobil as part of the latter's covert campaign to support organizations that challenge evidence of global warming, in an effort to subvert the scientific consensus on climate change (Krugman, "Enemy"; Moyers).

The pronouncements of James Inhofe, the fundamentalist senator representing Oklahoma, supply another instance of the unholy alliance between corporate capital and dominionist antienvironmentalism. Inhofe has publicly stated that global warming is "the greatest hoax ever perpetrated on the American people" and claimed that "global warming is a UN conspiracy" (Wilson, "US Senator"). Now, Inhofe may be genuinely skeptical about global warming (at best, a perverse position given the overwhelming evidence to the contrary). However, the histrionics and stridency of his talk about "hoax" and "conspiracy" cannot be divorced from the fact that he is the recipient of hundreds of thousands of dollars from the oil and gas industry. According to the

Center for Responsive Politics, which publishes for public scrutiny campaign finance contributions to congressmen as released by the Federal Election Commission, Inhofe, in the 1999–2004 Senate election cycle, received nearly $500,000 from the energy and natural resource sector, by far his largest source of funding (opensecrets.org). Furthermore, the magnitude of this contribution reflected his chief committee assignment in Washington—chair of the Senate Committee on Environmental and Public Works (2003–7).

In summary, as we review the statements of Burkett and Beisner, of Beliles and McDowell, of ICES and Inhofe, it is quite evident that dominionism is consonant with the efforts of successive administrations to deregulate environmental protection in the service of corporate capital.

Fundamentalist dominionism is certainly susceptible to a conventional ecological critique—that is to say, one framed in *scientific-environmentalist* terms of its unsustainability as a practice, given nature's finite resources and the fragility of ecosystems. The same critique would also respond to belief in the Rapture as an abdication from responsibility for the preservation of the earth and its species. Alternatively, a postmodern ecological critique has the conceptual resources to contest dominionism at the level of its *discursive* transactions; that is to say, the narrative frames and interpretive methods through which fundamentalists have constructed their understanding of the natural world. Here, I shall suggest how postmodernism enables critical standpoints that, collectively, open a second front in an engagement with the dominionist model of humanity's relationship to nature. These standpoints will be discussed under three rubrics: contingency, counter-transcendent thinking, and zoontology.[3]

Notes

1. Cizik has a formidable enemy in the person of James Dobson, proprietor (1977–2003) of the "Focus on the Family" media empire and, according to Dan Gilgoff, writing in 2007, the most powerful leader of the Christian Right. As the National Association of Evangelicals' chief Washington lobbyist, Cizik has advocated action on global warming, a cause that Dobson sees as a distraction from the evangelical moral agenda. Accordingly, Dobson sought, but failed, to have Cizik expelled from the association.
2. Norman Wirzba, a Christian environmentalist, has advanced his own critique of the rhetoric of stewardship used in the "Cornwall Declaration." Noting that "an economic agenda ... has taken the lead in defining the parameters of stewardship" (*Paradise* 130), he aims to redefine those parameters on what he believes to be an appropriately Christian basis:

> But the steward is not the outright owner of the house. He or she is trusted by someone else, presumably the owner of the house, to manage the affairs of the house in a manner that would be pleasing to the owner. A steward of creation, then, is someone who acknowledges God as the owner and master of the creation, yet recognizes himself or herself as entrusted with the creation's wise management and conservation, rather than its exploitation. (129)

Wirzba also reminds the Christian conservatives who endorsed the declaration's doctrine of "human mastery . . . without limit" (129) of the inversion of power relations promoted by the Gospels. The latter elevate servanthood above mastery. For example, Luke quotes Jesus: "I am among you as one who serves" (22:27 RSV; see also Phil. 2:7). And the mandate to "till and keep" the land (Gen. 2:15 RSV) imposes an ethic of service, not domination (139). The demeanor of servanthood guards against the human self-aggrandizement at nature's expense that could upset the ecological balance of the Creation.

3. Spiritual ecology can provide another kind of critique of dominionism. However, eco-spiritual movements are diverse; their ethics cannot be summarized under a single rubric. There are the spiritual resonances of Deep Ecology's ethic of "re-earthing," New Age eco-spiritualities that stress "cosmic interconnectedness," and eco-theologies, directly inspired by scriptural teachings about the Creation, as in the Christian environmentalism of Wirzba's *The Paradise of God*. Whether the thinking behind these movements qualifies as postmodern is debatable (see Maltby 61–71).

4. Metaphysical assumptions may linger undetected in postmodern thought; all the same, it can more easily (and modestly) face up to its contingencies and conjunctural limits than fundamentalism, whose grand narrative and piousness preclude it from adopting a self-reflexive or self-situating stance. Indeed, fundamentalism positions itself at, as it were, a trans-contextual viewpoint, beyond time and change, in claiming knowledge of the cosmic order or eternal redemption. Evidently, there is no place for the role of contingency in the fundamentalist commitment to the literal truth of the scriptures as the Word of God; inevitably, this belief entails faith in the absolute and necessary authority of the Bible.

5. In *The Seeds of Time*, Fredric Jameson discusses how our conceptions of "nature" reflect shift s in our hopes for or expectations of revolutionary change. While the modernist conception of nature as controllable—a "Promethean Utopianism" (48)—was symptomatic of faith in

the transformative power of revolutionary politics, the postmodernist conception of nature as a fragile ecosystem that requires from us a "self-policing attitude," a "new style of restraint" vis-à-vis our "collective ambitions" (48), reflects our loss of faith in the chances for revolutionary change in the late-capitalist age. Jameson writes, "It seems to be easier for us today to imagine the thorough-going deterioration of the earth and of nature than the breakdown of late capitalism" (xii). The free market has been naturalized to the point where we, in the liberal "democracies," cannot conceive of its collapse; hence, we now tend to figure collapse in biological terms. (And here we should note that, in the Left Behind novels, accounts of the apocalypse focus on the destruction of the biosphere rather than the socioeconomic sphere.) In short, for Jameson, ecological discourse, though it has a positive, utopian inflection, also serves an ideological function today, channeling our ideas of breakdown through biological rather than political images.

6. Yet, Nietzsche has also been cited to critique environmentalism in the name of a Judeo-Christian-inspired "green guilt." Stephen Asma finds in our zealous and dogmatic environmentalism the expression of an overdeveloped conscience, as bequeathed by Judeo-Christian values. Oppressed by guilt and a sense of unworthiness vis-à-vis the Creation, secular culture substitutes environmentalism for religion, pollution for sin, ecocide and global warming for apocalyptic scenarios. Recalling Nietzsche's premise that the more a Christianized humanity humbles itself, the more it exalts God, Asma writes: "We [now] need a belief in a pristine environment because we need to be cruel to ourselves as inferior beings, and we need that because we have these aggressive instincts that cannot be let out." Moreover, he frames his argument to encompass the secular liberal/Christian conservative divide: "The same demographic group for whom religion has little or no hold (namely white liberals) turns out to be the most virulent champions of all things green. Is it possible that these folks must vent their moral spleen on environmentalism because they don't have all the theological campaigns (e.g., opposing gay marriage, opposing abortion) on which social conservatives exercise *their* indignation?"

7. David Hume, in his critique of arguments from design for the existence of God, famously hypothesized: "This world . . . was only the first rude essay of some infant deity who afterwards abandoned it, ashamed of his lame performance; it is the work only of some dependent, inferior deity, and is the object of derision to his superiors; it is the production of old age and dotage in some superannuated deity, and ever since his death has run on at adventures, from the first impulse and active force which it received from him" (108).

8. Cheyney mobilizes his concept of "bioregional narrative" to resist the universalizing purview of modernist grand narratives, including those implicit in purely objective, science-based environmentalist panaceas. Here, I invoke the concept for its anti-transcendent character.

9. For Brian Campbell, bioregionalism "highlight[s] the potentially powerful bonds between people and places as a resource for environmentalism. We act ethically because we care about the place we call home. . . . Bioregionalism asserts that the most important response to the ecological crisis is to 'reinhabit' place" (207). Proponents of bio-regionalism see it as fostering a stable place-based identity and as nurturing a sustainable relationship with the environment. However, one strand of postmodern ecology is highly critical of place-based ethics insofar as it is seen as conducive to conservative ideologies. For example, Whitney Bauman argues that a place-based conception of nature reifies and provincializes identity, preserves social inequalities, favors private property, promotes an ecology of stasis, and diminishes our sense of planetary collectivity. Hence, he advocates an "ecological nomadism," whose ethics of mobility serves to counter the "reactionary" effects of a place-based sense of home and, in particular, to enhance our sense of the common grounds between peoples.

10. Recall Wittgenstein's (albeit pre-postmodern) proposition: "If a lion could talk, we could not understand him" (223).

11. In their monograph *Kafka: Toward a Minor Literature,* Gilles Deleuze and Félix Guattari have explored that zone of indeterminacy with a view to extolling not power over animals but power in "becoming animal." They propose that "to become animal is to . . . cross a threshold, to reach a continuum of intensities that are valuable only in themselves, to find a world of pure intensities where all forms come undone, as do all the significations, signifiers, and signifieds, to the benefit of an unformed matter of deterritorialized flux, of non-signifying signs" (13). Becoming-animal is a "creative line of escape" that "replaces subjectivity" (36). To be sure, this change constitutes but a respite from a territorialized/bureaucratized mind-set, insofar as the latter, Deleuze and Guattari insist, is certain to reassert itself. All the same, they argue that embracing the "intensities" of animality amounts to an invigorating, life-enhancing experience.

12. As a matter of etymological and theological interest, when Jesus speaks of hell, he uses the Greek word *Geenna* (Matt. 23:33). The word, which can be traced back to Hebrew, means "the valley of the son(s) of Hinnom" (Josh. 15:8; see also 2 Kings 23:10). This valley, located just outside Jerusalem's city walls, was a refuse dump, where fires continuously burned, consuming mounds of stinking trash. In short, Jesus figures hell as a garbage heap.

DISCUSSION QUESTIONS

1. What did Lynn White Junior explain about the ecological implications of the "creation" story in Genesis?
2. What is dominion theology?
3. What is the connection between anti-environmentalism and Christian fundamentalism? Give examples from the reading.
4. Break down E. Calvin Beisner's logical argument for dominion.
5. Explain the mission and development of Creation Care.
6. Give examples of fundamentalist dominion within American politics.

POLITICS

The majority religion In the United States is Christianity, and most Christian branches and denominations are represented in this country. About 40% of the American Christian population fall under various forms of Evangelical movements. The majority of Evangelical Christian Americans deny evolution and follow creation science. This same demographic also has powerful influence over the political process within the United States. Homosexuality is one topic that exemplifies the tension of this overlap of religion and politics. The following article explores this evolving relationship.

ARTICLE 2

HOMOSEXUALITY

By Mark A. Smith

As contending translations of *malakos* and *arsenokoitēs* and the disputes over how to understand the homosexual practices Paul described. Revisionist arguments may exert some persuasive influence, but the shift in Christian opinion primarily reflects the tenor of the times. On homosexuality as on other issues, religious believers usually follow the same trends as the rest of society.

The Softening Political Rhetoric of Christian Conservatives

Reflecting the values of a changing society, Christian leaders—including conservative evangelicals—have called for compassion and accepted certain aspects of gay rights. Even as they opposed one important right, same-sex marriage, evangelical leaders softened their political rhetoric. To understand this change, we can examine one of the earliest electoral battles over homosexuality, which occurred in Florida in Dade County (later renamed Miami-Dade County). In 1977 the county passed an ordinance prohibiting discrimination in employment or housing based on a person's sexual orientation. Dade County contained the city of Miami, home to a sizeable gay population, and supporters of gay rights viewed the ordinance as model legislation that other counties and cities might adopt.

What happened next dashed those lofty hopes. The Dade County ordinance provoked a countermovement led by Anita Bryant, a popular singer with four Top 40 singles to her name. Focused initially on a referendum to repeal the Dade County ordinance, Bryant's campaign soon galvanized nationwide resistance to gay rights. Explaining that she entered the political arena after prayer and conversations with her husband and her pastor, Bryant sprinkled her speeches and writings with references to God, Jesus, the Bible, and Christian morality. Christian conservatives rallied to her side, making her one of the first leaders of the movement subsequently labeled as the Christian right.[1] Bryant's denomination, the Southern Baptist Convention, praised her "courageous stand" in a resolution, and the group commended her yet again the following year.[2]

To coordinate her campaign, Bryant formed an organization called Save Our Children. The title was revealing, for it prompted those hearing it to ask: "From what, exactly?" Bryant explained her answer: "Homosexuals cannot reproduce—and so they must recruit. And to freshen their ranks, they must recruit the youth of America."[3] By endorsing a degenerate lifestyle, the Dade County ordinance therefore offered "an open door to homosexual recruitment." Children who resisted the siren's call of temptation could still suffer because, in Bryant's mind, homosexuals commonly engaged in pedophilia. The threat was especially intense in schools, where "a particularly deviant-minded teacher could sexually molest children."[4] To Bryant, this was no idle threat; she believed that homosexuals molested children all across America. People began sending her stories of sexual abuse that gays allegedly committed, and she told a reporter that these accounts "would turn your stomach."[5] In her view, and that of her supporters, Save Our Children captured in its title the reason why the gay rights movement must be stopped.

Within a couple of years, Bryant faded from the national scene after encountering personal and financial difficulties. With Bryant largely on the sidelines, others stepped forward to build the Christian right into a powerful movement. Rev. Jerry Falwell, cofounder of the Moral Majority in 1979, was arguably the most influential such leader among Christian conservatives in the 1980s. Throughout his long career in the public eye, Falwell attacked homosexuality and repudiated the lgbt movement's political agenda, famously blaming 9/11 partly on God's angry response to America's acceptance of the gay lifestyle.[6] Other leaders in the 1980s and 1990s, such as the Christian broadcaster and presidential candidate Pat Robertson, echoed Falwell in denouncing anything and everything associated with homosexuality.

By the second decade of the twenty-first century, however, leaders of the Christian right were using less divisive language on gay rights issues. For example, in 2010 the Family Research Council made an illuminating set of arguments in opposing the repeal of "Don't Ask, Don't Tell." In press releases and e-mail alerts to members, the frc asserted that overturning the Clinton-era policy would undermine military effectiveness.[7] Among its other arguments, the frc stated that Congress should focus instead on the economy and that top military officials opposed the new initiative.[8] But the frc declined to follow in Anita Bryant's footsteps by demonizing homosexuals. The frc's press releases and e-mail alerts never asserted or even implied that repealing "Don't Ask, Don't Tell" would allow homosexuals throughout society to recruit and sexually molest children. Whether intentionally or not, the frc treated gays and lesbians with much more respect than Bryant did.

Christian conservatives have also softened their rhetoric on the most visible and controversial gay rights issue of the twenty-first century: marriage. In 2007 prominent Christian intellectuals and organizers formed the National Organization for Marriage (NOM), which counted fighting same-sex marriage among its primary goals. Coordinating with supporters at the state and local levels, nom distributed information and developed strategies, giving its allies guidance, among other things, on the most effective means of political persuasion. In the section of its website called "Marriage Talking Points," which focused exclusively on same-sex marriage, nom stated, "Extensive and repeated polling agrees that the single most effective message is: 'Gays and lesbians have a right to live as they choose, they don't have the right to redefine marriage for all of us.' This allows people to express support for tolerance while opposing gay marriage."[9]

For many years opponents of same-sex marriage used this kind of language to explain their positions. While running for vice president on the Republican ticket in 2008, for example, Sarah Palin gave a revealing answer to a question about civil unions and gay marriage. She explained that she opposed extending civil unions beyond Alaska and into the rest of the nation

> if it goes closer and closer towards redefining the traditional definition of marriage between one man and one woman. And unfortunately that's sometimes where those steps lead. But I also want to clarify, if there's any kind of suggestion at all from my answer that I would be anything but tolerant of adults in America choosing their partners, choosing relationships that they deem best for themselves, you know, I am tolerant and I have a very diverse family and group of friends and even within that group you would see some who may not agree with me on this issue, some very dear friends who don't agree with me on this issue.[10]

In her answer Palin followed both elements of nom's messaging strategy—opposition to "redefining marriage" paired with support for "tolerance"—while avoiding once-common talking points involving deviant behavior, child molestation, and recruitment into the homosexual ranks.

Whether Palin's appeal for tolerance reflected her genuine sentiments or mere expediency, she used language undeniably more favorable to gays and lesbians than Anita Bryant and Jerry Falwell did in earlier decades. With certain exceptions, such as the campaign in 2008 to repeal a court decision for same-sex marriage in California, prominent Christian conservatives now steer clear of describing homosexuals as likely pedophiles. In short, Christian conservatives, through their public discourse, have accommodated modern attitudes about homosexuality. The need to develop and deploy political messages that resonate with Americans has encouraged Christian organizations like nom and political figures like Sarah Palin to incorporate a central element of the contemporary culture—tolerance—into their rhetoric even as they opposed extending tolerance into support for same-sex marriage.

As of 2013 Christian conservatives had succeeded in preventing same-sex marriage from spreading to the entire country. Over a longer period, however, NOM's poll-tested strategy of political messaging may confront its natural limit. If homosexuals are perverts who deserve criminal punishment or psychiatric treatment, as Americans assumed for much of the twentieth century, then it is obvious why marriage laws should treat such people differently. If, on the other hand, homosexuals are "choosing their partners, choosing relationships that they deem best for themselves," as Sarah Palin explained, it becomes harder to claim that they should not have the right to marry the person of their choice. By showing respect and promoting tolerance, evangelical leaders and their political allies may have helped undermine the policy goals they seek to achieve.

Catholic leaders face a similar dilemma. Responding in 2013 to a question about homosexuality, Pope Francis famously answered, "If a person is gay and seeks the Lord and has good will, who am I to judge that person?"[11] According to commentators around the world, the pope had moved the Catholic Church toward greater acceptance of gays and lesbians.[12] His full statements on this and other occasions, however, contain considerable nuance. Official Catholic teachings, which he explicitly affirmed, have long distinguished between sexual orientation and sexual behavior. While Pope Francis refused to condemn people with same-sex attractions, he embraced the doctrine holding that they sin if they act upon their desires.[13] As we saw in figure 5.1, many ordinary Catholics in America reject this distinction, believing that homosexual relations are "not wrong at all." By advocating tolerance for people with the orientation, the pope ended up promoting tolerance for those who engage in the associated behaviors.

Popes rarely change the Church's official positions, but—as this episode shows—they can accommodate the modern world through the ways they explain those positions. On another occasion in 2013, Pope Francis referred to homosexuality and related issues in saying, "It is not necessary to talk about these issues all the time."[14] Whereas some Protestant groups have actually revised their teachings about the morality of homosexual relationships, such a move is unlikely in the Catholic Church. Institutionally committed to tradition and authority, the Catholic Church pays a great price when it openly overturns a long-standing doctrine. Church leaders can nevertheless adapt to cultural trends by reformulating their rhetoric. Future popes and bishops will likely follow in Pope Francis's footsteps by talking less about homosexuality and by taking a softer tone on the issue.

The Future of Gay Rights in America

What does contemporary politics portend for homosexuality as a political issue? Polling data confirms that young people are far more liberal than their elders on this subject. In the General Social Survey I use throughout this chapter, young adults (defined to include ages eighteen to twenty-nine) were more liberal on homosexuality than the rest of the population in each year from 1973 to 2012. Given these generational differences, the natural cycles of birth, maturation, and death create steady changes in the views of the overall public. Every year a small percentage of middle-aged and elderly people die, and they are replaced in the voting population by new eighteen-year-olds. This turnover in the population ensures that the liberalizing trends in public opinion continue.

These trends cannot be stabilized or reversed unless people become more conservative on homosexuality as they age. As twentysomethings mature, start careers, get married, raise children, and eventually welcome grandchildren into the world, perhaps they will come to reject the views about homosexuality that characterized their younger years. An age gap would still remain, with young people holding more favorable attitudes than their elders, but the aging process would dampen the liberalizing momentum within the general population.

It turns out, however, that on this issue, Americans actually tend to become more liberal, not more conservative, as they age. This effect appeared with the first cohort of young adults in the gss surveys, those aged eighteen to twenty-nine in 1973. Using subsequent gss data through 2012, we can track their attitudes as they reached their thirties, forties, fifties, and sixties. On questions of civil liberties for homosexuals, the cohort of young adults in 1973 held steady in their levels of support until 1988, when they started becoming more liberal. In 2012, now aged fifty-seven to sixty-eight, the cohort supported civil liberties for homosexuals at a rate of 67 percent, higher than the 59 percent they expressed four decades earlier as eighteen-to twenty-nine-year-olds. On the related question of morality, this cohort became more conservative in the 1980s before turning far more liberal in the 1990s and the first decade of the twenty-first century. In 2012, 35 percent of people aged fifty-seven to sixty-eight said homosexual relations were "not wrong at all," whereas only 20 percent of the cohort took that view back in 1973. Incidentally, this finding generalizes well beyond the people who were young adults in 1973. The gss data show that in 2012 every age cohort (those in their thirties, forties, etc.) held more liberal attitudes about homosexuality (for both civil liberties and morality) than those same cohorts did in their younger years.

Of course, the tendency for Americans to become more liberal on homosexuality with age does not change the fact that at any single point in time, younger people are much more liberal than older people. In addition, the younger Americans entering the adult population increasingly take a live-and-let-live attitude on homosexuality. Each year's set of eighteen-to twenty-nine-year-olds from 1993 to 2012 held more liberal attitudes (again, on both civil liberties and morality) than earlier groups did at the same age. As a result, the last group of young adults in 2012 was the most supportive on record.

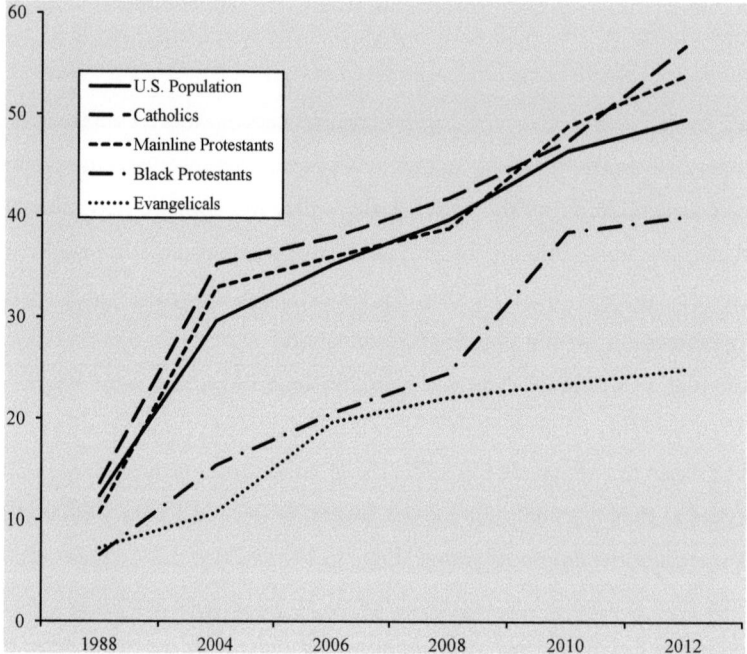

Figure 5.1. Percentage supporting same-sex marriage *Source:* General Social Survey

Reflecting the changes in public opinion, the gss eventually expanded its range of questions. For the first time in 1988 and then regularly since 2004, the gss solicited Americans' views about same-sex marriage. Interviewers asked respondents whether "homosexual couples should have the right to marry one another." Figure 5.1 plots the percentage of the general public, as well as four Christian subgroups, who "strongly agree" or "agree" that homosexual couples should have equal marriage rights.

In 1988, as figure 4.6 indicates, only 12 percent of the American public wanted to legalize same-sex marriage. That figure rose to 30 percent in 2004 and 50 percent in 2012—a fourfold increase in about two decades, albeit from a low starting point. Among the major Christian groups, the pattern is familiar by now. In each year captured on the graph, Catholics and mainline Protestants resemble the overall US population. Evangelical Protestants and black Protestants consistently express lower levels of support, but both groups' numbers trend upward with the rest of the American population. It is noteworthy that a greater share of evangelicals (25 percent) and black Protestants (40 percent) endorsed gay marriage in 2012 than the general public did (12 percent) in 1988.

Paralleling other public opinion questions relating to homosexuality, the American public divides by age. In 1988 only 14 percent of people aged eighteen to twenty-nine wanted to legalize same-sex marriage. Twenty-four years later in 2012, when those same people were now aged forty-two to fifty-three, their level of support had risen to 52 percent. Meanwhile, the new group of young adults in 2012 approved of gay marriage at an even higher rate of 65 percent. With population replacement compounding the effects of people growing more liberal as they age, public support for same-sex marriage continued to rise in the following years.

How will Christian leaders respond as the views of their members and the broader society keep changing? Will Christian leaders coming of age in a society with much less animus toward homosexuals develop attitudes and beliefs different from their predecessors? As Yogi Berra once said, "It's tough to make predictions, especially about the future." For the case at hand, we can nevertheless use the history of similar issues, of which divorce might be the closest parallel, to guess what will happen with homosexuality. Both divorce and homosexuality relate to family values and involve personal morality. The two issues, too, have each attracted attention and lobbying from Christian groups in different periods.

Over the course of American history, the public has altered its judgments of both issues. From the founding of the American colonies through much of the twentieth century, local communities frequently regarded divorcées as moral degenerates. As marriages fractured at increasing rates in the nineteenth and especially the twentieth centuries, divorce gradually lost its social stigma. A similar pattern has happened with homosexuality. In the not-too-distant past, gays and lesbians commonly kept their sexual orientations private to avoid jeopardizing their relationships with employers, family members, and heterosexual friends. In today's dramatically different environment, growing numbers of straight people accept gays and lesbians, especially those they know personally. The norm of tolerance, in fact, has spread so widely that conservative politicians and Christian leaders tout their own tolerance.

As with divorce, Christians have modified their views of homosexuality and adjusted their political stances. On a literal reading, the Bible seems clear in forbidding both divorce (except possibly on grounds of a partner's adultery) and homosexual acts. Indeed, many Christian groups historically took a hard line on both issues and pressed their governments to protect the moral order through restrictive laws. Modern scholars and advocates, however, often interpret the Bible as more open to divorce and homosexuality than it first appears. In both cases, the reinterpretations emerged within cultural environments that increasingly rejected the stringent legal requirements and societal attitudes of earlier times.

For the most part, ordinary Christians do not form their views by engaging modern biblical interpretations at a deep level. In the more common process, Christians bring to their religious communities the norms and values they absorb from the larger society, with the new biblical interpretations following and reinforcing rather than preceding and causing the changing attitudes. Several generations ago, churches made the then-controversial decision to welcome divorced people into their congregations. In the late twentieth and early twenty-first centuries, some churches made a similar move in welcoming homosexuals as members in good standing.

In an obvious difference between the two issues, the long-term interplay between public opinion, religious accommodation, and changes in public policy is complete (more or less) for divorce. The same point could be made for slavery, an issue that Christian groups debated in the eighteenth and nineteenth centuries. On homosexuality, by contrast, the transformations are ongoing and people are still divided. Roughly half of the American public backed same-sex marriage in 2012 and believed homosexual relations were "not wrong at all"; the other half disagreed. Given the current rates at which attitudes are changing, it might take until 2020 or 2030 before an overwhelming majority (say, 75 percent) of Americans takes the progay positions. As the levels of support increase, we can expect that many more states will legalize same-sex marriage. The US Supreme Court, meanwhile, could make a decision legalizing same-sex marriage nationwide. Indeed, at the beginning of its 2014–2015 term, the Supreme Court accepted on appeal a case that could lead to a landmark change in the constitutionality of bans on same-sex marriage.

How will organizations that represent Christian conservatives react to a situation where most or all states allow gay marriage? Again, the history of divorce may provide some insights. Once divorce became common and Christians began opposing strict laws regulating it, their leaders backed away from divorce as a political issue. This outcome does not mean that Christian leaders now approve of divorce; to the contrary—as explained in the previous chapter—many of them still call it a blight on society and a threat to children's well-being. But those same leaders bowed to the reality of divorce in American society and among their members. Lacking a constituency to limit divorce, Christian leaders removed the issue from their list of political priorities. Divorce became a private issue handled by individual families and churches rather than a public issue marked by intense lobbying, voter mobilization, and political conflict.

If public opinion continues along its current trend, the issue of homosexuality will wind up in the same place. Such a result would not require all or even most Christian leaders to embrace homosexual behavior as normal and legitimate. Instead those who deem it sinful would merely need to treat the issue, like divorce, as a question for private morality rather than public policy. They would give up the fight to repeal same-sex marriage in individual states or the country at large. Christian leaders and affiliated organizations would deemphasize homosexuality as a staple of fundraising appeals, campaign alerts, mass rallies, and other forms of political engagement, and they would no longer use the issue to determine which candidates to endorse for public office.

A development of this kind may sound like mere speculation. The same would have been said of someone who in 1890 predicted that Christian leaders of future generations would downplay divorce as a political issue. In the first half of the twentieth century, no one expected that one day homosexuals would come out of the closet, engage in sexual relations without fearing arrest, see favorable portrayals of themselves in movies and television shows, and gain the right to serve in the military. Compared to the massive changes in attitudes, beliefs, institutions, and laws that have already occurred, it is not far-fetched to anticipate a future in which Christian leaders either embrace same-sex marriage outright, as some have already done, or accommodate it by redirecting their political efforts elsewhere.

Notes

1. Volumes 1 to 5, covering 1900 to 1928, filled 15,734 pages. Using a random sample from throughout the period, I found that each page indexed an average of seventy-five articles, leading to a total of more than one million articles indexed from 1900 to 1928. Although some of those articles may have referred to homosexuality in passing, the compilers of the *Readers' Guide* did not perceive homosexuality as the central subject of any of them.
2. *Readers' Guide to Periodical Literature* 20 (March 1955 to February 1957): 1144–45.
3. John D. Emilio, *Sexual Politics, Sexual Communities: The Making of a Homosexual Minority in the United States*, 1940–1970 (Chicago: University of Chicago Press, 1983), 41–51.
4. As quoted in Lillian Faderman, *Odd Girls and Twilight Lovers: A History of Lesbian Life in Twentieth-Century America* (New York: Columbia University Press, 1991), 164.

5. Allan Berube, *Coming Out Under Fire: The History of Gay Men and Women in World War II* (New York: Free Press, 1990).

6. US Senate, "Employment of Homosexuals and Other Sex Perverts in the US Government," reprinted (selections) in *We Are Everywhere: A Historical Source-book of Gay and Lesbian Politics*, ed. Mark Blasius and Shane Phelan (New York: Routledge, 1997), 241–51 (quote is from p. 251).

7. David K. Johnson, *The Lavender Scare: The Cold War Persecution of Gays and Lesbians in the Federal Government* (Chicago: University of Chicago Press, 2004), 166–69.

8. American Psychiatric Association, *Diagnostic and Statistical Manual: Mental Disorders* (Washington, DC: American Psychiatric Association, 1952), 38–39.

9. Herb Kutchins and Stuart A. Kirk, *Making Us Crazy: DSM; The Psychiatric Bible and the Creation of Mental Disorders* (New York: Free Press, 1997), 58–60.

10. Founded in Chicago in 1924, the short-lived Society for Human Rights preceded the Mattachine Society as the earliest known gay rights organization in America.

11. Mattachine Society, "Missions & Purposes," reprinted (selections) in *Homosexuals Today: A Handbook of Organizations & Publications*, ed. Marvin Cutler (Los Angeles: One, 1956), 13–14.

12. Daughters of Bilitis, "Statement of Purpose" [1955], reprinted in Blasius and Phelan, *We Are Everywhere*, 328.

13. Figure based on word searches of the *New York Times* archive, accessed November 4, 2011, http://www.nytimes.com/ref/membercenter/nytarchive.html.

14. Gary J. Gates, "How Many People Are Lesbian, Gay, Bisexual, and Trans-gender?" Williams Institute, ucla School of Law, April 2011, accessed December 2, 2014, http://williamsinstitute.law.ucla.edu/wp-content/uploads/Gates-How-Many-People-LGBT-Apr-2011.pdf.

15. Hazel Gaudet Erskine, "The Polls: Morality," *Public Opinion Quarterly* 30 (Winter 1966–1967): 680. When reporting the results, Louis Harris and Associates apparently grouped the "don't knows" (i.e., the people unable to answer the question) with the people giving the substantive answer that homosexuals "don't help or harm" America. I have thus calculated the percentage saying "don't help or harm" by subtracting an estimate of 9 percent for the "don't knows," which is the figure yielded by an identical Louis Harris and Associates survey five years later.

16. American Psychiatric Association, *Diagnostic and Statistical Manual: Mental Disorders*, 2nd ed. (Washington, DC: American Psychiatric Association, 1968), 41–46.

17. Martin Duberman, *Stonewall* (New York: Plume, 1994).
18. David Carter, *Stonewall: The Riots that Sparked the Gay Revolution* (New York: St. Martin's Press, 2004).
19. Alan Yang, "The Polls—Trends: Attitudes toward Homosexuality," *Public Opinion Quarterly* 61 (1997): 484.
20. Benjamin I. Page and Robert Y. Shapiro, *The Rational Public: Fifty Years of Trends in Americans' Policy Preferences* (Chicago: University of Chicago Press, 1992).
21. Evelyn Hooker, "The Adjustment of the Male Overt Homosexual," *Journal of Projective Techniques* 21 (1957): 18–31.
22. Judd Marmor, ed., *Sexual Inversion: The Multiple Roots of Homosexuality* (New York: Basic Books, 1965).
23. Kutchins and Kirk, *Making Us Crazy*, 67.
24. Ronald Bayer, *Homosexuality and American Psychiatry: The Politics of Diagnosis* (New York: Basic Books, 1981).
25. Robert L. Spitzer, "A Proposal about Homosexuality and the apa Nomenclature: Homosexuality as an Irregular Form of Sexual Behavior and Sexual Orientation Disturbance as a Psychiatric Disorder," *American Journal of Psychiatry* 130 (1973): 1214–16.
26. Simon LeVay, "A Difference in Hypothalmic Structure between Heterosexual and Homosexual Men," *Science* 253 (August 1991): 1034–37.
27. J. M. Bailey, M. P. Dunne, and N. G. Martin. "Genetic and Environmental Influences on Sexual Orientation and Its Correlates in an Australian Twin Sample," *Journal of Personality and Social Psychology* 78 (March 2000): 524–36.
28. R. Blanchard and P. Klassen, "H-Y Antigen and Homosexuality in Men," *Journal of Theoretical Biology* 185 (April 1997): 373–78.
29. "The Homosexual in America," *Time*, January 21, 1966, 41.
30. W. Lance Bennett, "Toward a Theory of Press-State Relations in the US," *Journal of Communication* 40 (Spring 1990): 103–25; John R. Zaller, *The Nature and Origins of Mass Opinion* (New York: Cambridge University Press, 1992).
31. Edward Alwood, *Straight News: Gays, Lesbians, and the News Media* (New York: Columbia University Press, 1996).
32. Tom W. Smith, Peter Marsden, Michael Hout, and Jibum Kim, *General Social Surveys, 1972–2012* [machine-readable data file] (Chicago: National Opinion Research Center, 2013).

33. This and all subsequent references to the General Social Survey are based on my own analyses of the raw data.
34. This figure is essentially a mirror image of one showing the percentage of people saying "always wrong."
35. In most of the graphs I present throughout the book, I use a method called "locally weighted scatterplot smoothing" (LOESS). Measures of variables over time contain both real change—of interest here—and random variations from year to year owing to factors such as sampling error. loess allows the analyst to smooth out the short-term fluctuations so that the eye can focus on the long-term trends. As an added benefit, which is relevant later in the chapter, loess allows me to present the figures from multiple religious groups on the same graph. Without this smoothing, some of my graphs would be too cluttered to read.
36. Donald P. Haider-Markel and Mark R. Joslyn, "Beliefs about the Origins of Homosexuality and Support for Gay Rights: An Empirical Test of Attribution Theory," *Public Opinion Quarterly* 72 (2008): 291–310; Gregory B. Lewis, "Does Believing Homosexuality Is Innate Increase Support for Gay Rights?" Policy Studies Journal 37 (November 2009): 669–93.
37. Gregory M. Herek and John P. Capitanio, "'Some of My Best Friends': Inter-group Contact, Concealable Stigma, and Heterosexuals' Attitudes toward Gay Men and Lesbians," *Personality and Social Psychology Bulletin* 22 (1996): 412–24; Paul R. Brewer, *Value War: Public Opinion and the Politics of Gay Rights* (Lanham, MD: Rowman & Littlefield, 2008), ch. 3.
38. Robert Andersen and Tina Fetner, "Cohort Differences in Tolerance of Homosexuality: Attitudinal Change in Canada and the United States, 1981–2000," *Public Opinion Quarterly* 72 (Summer 2008): 311–30.
39. William N. Eskridge Jr., *Dishonorable Passions: Sodomy Laws in America*, 1861–2003 (New York: Viking, 2008).
40. Gary Mucciaroni, *Same Sex, Different Politics: Success and Failure in the Struggles over Gay Rights* (Chicago: University of Chicago Press, 2008), ch. 4.
41. "Employment Non-discrimination Laws on Sexual Orientation and Gender Identity," Human Rights Campaign, accessed March 21, 2013, http://preview.hrc.org/issues/4844.htm.
42. "Fair Housing Laws: Renters' Protection from Sexual Orientation Discrimination." FindLaw, accessed March 21, 2013, http://civilrights.findlaw.com/discrimination/fair-housing-laws-renters-protection-from-sexual-orientation.html.
43. Margot Canaday, *The Straight State: Sexuality and Citizenship in Twentieth-Century America* (Princeton, NJ: Princeton University Press, 2009), ch. 5.

44. Chad C. Carter and Anyony Barone Kolenc, "'Don't Ask, Don't Tell': Has the Policy Met Its Goals?" *University of Dayton Law Review* 31 (2005): 1–24.
45. See, for example, "A Quiet End," *Mobile Register*, December 27, 2010, A5.
46. The only exception came in Arizona in 2006, when voters rejected a constitutional amendment that would have banned both same-sex marriage and civil unions. In 2008 Arizona voters passed a constitutional amendment that banned only same-sex marriage.
47. "Same-Sex Marriage Laws." National Conference of State Legislatures, accessed November 23, 2014, http://www.ncsl.org/research/human-services/same-sex-marriage-laws.aspx.
48. The Church of England, which released its influential Wolfenden Report in 1957, acted a full decade before any American denomination. Proceeding without the official sanction of a national body, a group of British Quakers in 1963 also engaged the issue of homosexuality several years before any American denominations. See J. Gordon Melton, *The Churches Speak on Homosexuality: Official Statements from Religious Bodies and Ecumenical Organizations* (Detroit: Gale Research, 1991), 57–65 and 181–200.
49. Augustine, *Confessions*, trans. Henry Chadwick (Oxford: Oxford University Press, 1991), 3:8:15.
50. Timothy M. Renick, *Aquinas for Armchair Theologians* (Louisville, KY: West-minster John Knox Press, 2002), 87–88.
51. Sabina Flanagan, *Hildegard of Bingen: A Visionary Life* (New York: Routledge, 1999), 66–67.
52. As quoted in Ewald Martin Plass, *What Luther Says: An Anthology* (St. Louis: Concordia Publishing House, 1959), 1:134.
53. "Westminster Catechism," reprinted (selections) in Melton, *The Churches Speak on Homosexuality*, 146.
54. United Methodist Church, "Resolution on Health, Welfare, and Human Development," reprinted (selections) in Melton, *The Churches Speak on Homosexuality*, 240.
55. United Methodist Church, "Social Principles," reprinted (selections) in Melton, *The Churches Speak on Homosexuality*, 241.
56. Unfortunately, the 1970 survey did not ask detailed questions about the respondent's religious affiliation, making it impossible to know for certain how United Methodists compared to the general population.
57. Council for Christian Social Action of the United Church of Christ, "Resolution on Homosexuals and the Law," reprinted in Melton, *The Churches Speak on Homosexuality*, 203–4.

58. Wendy Cadge, "Vital Conflicts: The Mainline Denominations Debate Homosexuality," in *The Quiet Hand of God: Faith-Based Activism and the Public Role of Mainline Protestantism*, ed. Robert Wuthnow and John H. Evans (Berkeley: University of California Press, 2002), 272–73.

59. United Presbyterian Church in the United States of America, "Sexuality and the Human Community," reprinted (selections) in Melton, *The Churches Speak on Homosexuality*, 147–48.

60. Lutheran Church in America, "Sex, Marriage, and Family," reprinted (selections) in Melton, *The Churches Speak on Homosexuality*, 113.

61. Southern Baptist Convention, "On Homosexuality," reprinted in Melton, The Churches Speak on Homosexuality, 200.

62. The complete set of resolutions passed by the Southern Baptist Convention can be found at the group's website (http://www.sbc.net/resolutions/default.asp, accessed April 20, 2011).

63. National Conference of Catholic Bishops, "Principles to Guide Confessors in Questions of Homosexuality," reprinted in Melton, *The Churches Speak on Homosexuality*, 2–9.

64. To classify people as Catholics, mainline Protestants, evangelical Protestants, and black Protestants, I use the system developed by Brian Steensland and his collaborators. (I used the same system in my previous chapter on divorce.) See Brian Steensland et al., "The Measure of American Religion: Toward Improving the State of the Art," *Social Forces* 79, no. 1 (September 2000): 291–318.

65. Congregation for Catholic Education, "Instruction Concerning the Criteria for the Discernment of Vocations with regard to Persons with Homosexual Tendencies in View of Their Admission to the Seminary and to Holy Orders," The Vatican, accessed April 20, 2011, http://www.vatican.va/roman_curia/congregations/ccatheduc/documents/rc_con_ccatheduc_doc_20051104_istruzione_en.html.

66. *The Book of Discipline of the United Methodist Church* (Nashville, TN: Abingdon Press, 2009), par. 2702.1.

67. Jerald Hyche and Pat McCaughan, "Bishops Affirm Openness of Ordination Process," Episcopal News Service, July 14, 2009, Episcopal Church, accessed November 23, 2014, http://library.episcopalchurch.org/sites/default/files/daily07_071409.pdf; "elca Assembly Opens Ministry to Partnered Gay and Lesbian Lutherans," *The Lutheran*, August 21, 2009, Evangelical Lutheran Church in America, accessed November 23, 2014, http://www.thelutheran.org/blog/index.cfm?person_id=194&blog_id=1324.

68. "Presbyterian Church (USA) Approves Change in Ordination Standard," Presbyterian Church (USA), accessed May 11, 2011, http://www.pcusa.org/news/2011/5/10/presbyterian-church-us-approves-change-ordination.

69. "On Homosexuality and the United States Military" [June 2010] and "On Biblical Sexuality and Public Policy" [June 2009], both available at the Southern Baptist Convention's website (http://www.sbc.net/resolutions/default.asp, accessed June 12, 2011).

70. Dave Bohon, "Black Pastors Challenge naacp's Support for Same-Sex Marriage," *New American*, July 17, 2012, accessed July 19, 2012, http://www.thenewamerican.com/culture/faith-and-morals/item/12094-black-pastors-challenge-naacppercentE2percent80percent99s-support-for-same-sex-marriage.

71. "Churchgoers Disapprove of Gay and Lesbian Pastors," Rasmussen Reports, June 30, 2006, accessed April 20, 2011, http://www.rasmussenreports.com/public_content/politics/general_politics/june_2006/churchgoers_disapprove_of_gay_and_lesbian_pastors.

72. I identified the titles through WorldCat, a database that compiles the collections of 70,000 libraries around the world. For the subjects under investigation here, books published in the United States constitute a clear majority. The data in figure 4.5 include all books the Library of Congress classifies under the subjects of "homosexuality—religious aspects—Christianity," "homosexuality—religious aspects—biblical teaching," "homosexuality in the Bible," and "homosexuality—biblical teaching." Most of the titles are traditional books bearing the imprint of commercial, university, or nonprofit presses, but some are reports from advocacy groups, denominations, and other religious bodies.

73. Marion L. Soards, *Scripture and Homosexuality: Biblical Authority and the Church Today* (Louisville, KY: Westminster John Know Press, 1995), 24.

74. Bruce Hilton, *Can Homophobia Be Cured? Wrestling with Questions that Challenge the Church* (Nashville, TN: Abingdon Press, 1992), 75.

75. Joe Dallas, *A Strong Delusion: Confronting the "Gay Christian" Movement* (Eugene, OR: Harvest House, 1996), 187–88.

76. Victor Paul Furnish, "What Does the Bible Say about Homosexuality?" in *Caught in the Crossfire: Helping Christians Debate Homosexuality*, ed. Sally B. Geis and Donald E. Messer (Nashville, TN: Abingdon Press, 1994), 59.

77. Soards, *Scripture and Homosexuality*, 16.

78. Dallas, *A Strong Delusion*, 191–92. Biblical quotations in this chapter are taken from *The Holy Bible, New Revised Standard Version* (Grand Rapids, MI: Zondervan, 1989).

79. Dallas, *A Strong Delusion*, 92–93.
80. Hilton, *Can Homophobia Be Cured?*, 70.
81. James B. De Young, *Homosexuality: Contemporary Claims Examined in Light of the Bible and Other Ancient Literature and Law* (Grand Rapids, MI: Kregel, 2000), 189–95.
82. Dallas, *A Strong Delusion*, 198.
83. Furnish, "What Does the Bible Say about Homosexuality?," 61; John Bos-well, *Christianity, Social Tolerance, and Homosexuality: Gay People in Western Europe from the Beginning of the Christian Era to the Fourteenth Century* (Chicago: University of Chicago Press, 1980), 341–53.
84. Robin Scroggs, *The New Testament and Homosexuality* (Philadelphia: Fortress Press, 1983), 107–109.
85. De Young, *Homosexuality*, 198–99.
86. Dallas, *A Strong Delusion*, 194.
87. Soards, *Scripture and Homosexuality*, 20–23.
88. Victor Paul Furnish, *The Moral Teachings of Paul*, 2nd ed. (Nashville, TN: Abington Press, 1985), 52–82.
89. Boswell, *Christianity, Social Tolerance, and Homosexuality*, 109.
90. Stanley J. Grenz, *Welcoming but Not Affirming: An Evangelical Response to Homosexuality* (Louisville, KY: Westminster John Knox Press, 1998), 60.
91. Tom Horner, *Jonathan Loved David: Homosexuality in Biblical Times* (Philadelphia: Westminster Press, 1979).
92. Hilton, *Can Homophobia Be Cured?*, 75–76.
93. Jeffrey S. Siker, "Gentile Wheat and Homosexual Christians: New Testament Directions for the Heterosexual Church," in *Biblical Ethics & Homosexuality: Listening to Scripture*, ed. Robert L. Brawley (Louisville, KY: Westminster John Knox Press), 145–50.
94. Dallas, *A Strong Delusion*, 203–206.
95. The most obvious example is Troy Perry, founder of the Metropolitan Community Church. Perry has written several books that discuss, with varying degrees of detail, questions of the Bible and homosexuality. Other examples include David Day, *Things They Never Told You in Sunday School: A Primer for the Christian Homosexual* (Austin: Liberty Press, 1987); and Samuel Kader, *Openly Gay, Openly Christian: How the Bible Really Is Gay Friendly* (San Francisco: Leyland, 1999).
96. Siker, "Gentile Wheat and Homosexual Christians," 146; Paul R. Smith, *The Bible and Homosexuality: Affirming All Sexual Orientations as Gifts from God* (Kansas City, MO: Paul Smith, 1998), 28.

97. Jack Bartlett Rogers, *Jesus, the Bible, and Homosexuality: Explode the Myths, Heal the Church* (Louisville, KY: Westminster John Knox Press, 2006), xviii.
98. Barry D. Adam, *The Rise of a Gay and Lesbian Movement*, rev. ed. (New York: Twayne, 1995), ch. 6.
99. "Resolution on Homosexuality" [June 1977], Southern Baptist Convention, accessed August 4, 2011, http://www.sbc.net/resolutions/amResolution.asp?ID=607; "Resolution on Commendation of Anita Bryant" [June 1978], Southern Baptist Convention, accessed August 4, 2011, http://www.sbc.net/resolutions/amResolution.asp?ID=744.
100. Anita Bryant, *The Anita Bryant Story: The Survival of Our Nation's Families and the Threat of Militant Homosexuality* (Old Tappan, NJ: Fleming H. Revell, 1977), 62.
101. Ibid., 78, 114.
102. Morton Kondracke, "Anita Bryant Is Mad about Gays," *New Republic*, May 7, 1977, 13–14.
103. John Kenneth White, "Terrorism and the Remaking of American Politics," in *The Politics of Terror: The US Response to 9/11*, ed. William Crotty (Boston: Northeastern University Press, 2004), 55.
104. Family Research Council, *Tony Perkins' Washington Update*, e-mail newsletter. The arguments appeared in the following editions: "Ignore Amos?" November 8, 2010; "gop Clips Senate's (Left) Wing," December 1, 2010; "dadt Defeat Doesn't Mean Conservative Surrender," December 20, 2010."
105. Family Research Council, *Tony Perkins' Washington Update*, e-mail newsletter. The arguments appeared in the following editions: "Harry Knuckles . . . Under," November 18, 2010; "Lame Ducks Talk Turkey," November 19, 2010; "Repeal Loses Appeal among Troops," November 30, 2010; and "Mullen: Don't Like 'Don't Ask?' Don't Serve!" December 2, 2010.
106. "Marriage Talking Points." National Organization for Marriage, accessed April 20, 2011, http://www.nationformarriage.org/site/c.omL2KeN0LzH/b.4475595/k.566A/Marriage_Talking_Points.htm.
107. Transcript of Vice-Presidential Debate at Washington University in St. Louis, October 2, 2008, The American Presidency Project, The University of California, Santa Barbara, accessed April 20, 2011, at http://www.presidency.ucsb.edu/ws/index.php?pid=84382#axzz1K5ts1oNv.
108. A full transcript of Pope Francis's remarks can be found in "A Memorable In-Flight Press Conference," *Inside the Vatican*, August 2013, accessed November 4, 2013, https://insidethevatican.com/lead-story/a-memorable-in-flight-press-conference.

109. See, for example, "Pope Francis: Who Am I to Judge Gay People," bbc News, July 29, 2013, accessed November 4, 2013, http://www.bbc.co.uk/news/world-europe-23489702.

110. Antonio Spadaro, S.J., "A Big Heart Open to God: The Exclusive Interview with Pope Francis," *America: The National Catholic Review*, September 30, 2013, accessed November 4, 2013, http://www.americamagazine.org/pope-interview.

111. Ibid.

DISCUSSION QUESTIONS

1. Who were Anita Bryant and the Reverend Jerry Falwell? What did they believe about homosexuality? How were SaveOurChildren and the Moral Majority a part of their vision?
2. Give examples of what the author means by a "softening Christian rhetoric" about homosexuality,
3. Explain Pope Francis's remarks about homosexuality.
4. As Americans age, do they become more liberal or more conservative? Why do you think this is so?
5. Analyze the findings of support for same-sex marriage by the Christian branches as presented in Figure 4.6 of the article.
6. What can the relationship between divorce and homosexual marriage imply for the future?

WOMEN

Evangelical Christianity is a leader in abstinence-only sex education for children across the United States. Many private religious schools do not teach sexual education classes, and research shows that this lack of education usually leads to higher numbers of teenage pregnancies. While the Evangelical Christian community is coming to terms with how to deal with sexual education, many groups have taken it upon themselves to create websites that address sexual education for adults. More importantly, they are teaching married Christian women how to reach orgasm. The following article explores these websites.

ARTICLE 3

SEXUAL AWAKENING: DEFINING WOMEN'S PLEASURES

By Kelsy Burke

CarrieForChrist firmly believed that, as a married woman, God allowed—even required—her to enjoy sex with her husband. But starting on her wedding night, sexual intercourse was "extremely painful." She knew it wasn't *supposed* to be, but she did not know how to enjoy it, having only learned of the perils of sex from her evangelical Christian family, friends, and church. "The way I grew up, you didn't talk about sex," she told me. "You know, the old 'sex is bad' or taboo. I never got 'The Talk.'" Carrie didn't pursue information about sex for fear that what she found would offer ungodly advice; if it didn't come "from a faith-based perspective, it'd lead to confusion." And so she entered her marriage knowing very little about her sexuality. She confided to me, "I didn't know zilch about how my body worked down there before I got married—well, not counting the cycle every month ☺." The playful smiley face emoticon transfers the candid and intimate nature of women's conversations on Christian sexuality websites to our interview—women on these sites are, Carrie told me, honest, unpretentious, and friendly.

CarrieForChrist learned about LustyChristianLadies.com from her younger sister, whom Carrie describes as more "in touch" with her body, even though she's not yet married or sexually active. Carrie spent weeks carefully exploring the interactive blog site after first discovering it. She began to follow the routine daily posts. On Mondays, the website posts a weekly poll to LCL readers with a question like, "What's your favorite time of day to have sex?" On Tuesdays, there is a "task" for readers to accomplish that week, such as, "Leave a series of notes for him to find, all starting with 'I love your ...' Make some of them serious and some of them steamy!" On Thursdays, one of the LCL bloggers publishes a commentary about some topic related to sexuality, often prompted by a reader's question to the blog team. On Fridays, the site publishes various sentences related to sexuality and marriage, such as "The smell of ___ is a turn on for me!" and readers are asked to fill in the blank. They reply with comments like "Men's cologne," "His beautiful man parts!" and "Jasmine vanilla massage oil."

From this online dialogue, CarrieForChrist learned from other Christian women who loved sex and loved to talk about it. She read about practical tips to ease the pain she experienced during intercourse and got advice about ways to increase her pleasure, like by touching herself during sex with her husband. LCL bloggers and readers also convinced her that she shouldn't feel ashamed or embarrassed about giving or receiving oral sex, activities that appealed to CarrieForChrist but also gave her anxiety. "I remember one of the Tuesday tasks was something along the lines of 'surprise your hubby with something,' and I timidly put in a comment that I wanted to have the courage to give my husband a BJ [blow job]. Some of the comments were like, 'You can do it, girl!' And after I did it and LOVED it, I went back to that post and commented, 'it was WILD!'"

LustyChristianLadies.com helped CarrieForChrist realize her sexual potential and understand that she could be confident sexually and enjoy having sex with her husband. "It was encouraging to know that I wasn't the only one having difficulty," she told me. Carrie learned to overcome physical obstacles related to the pain she felt during intercourse, to overcome emotional hurdles of shame and embarrassment that she felt about sex, and to amend her belief system to incorporate religious values that encourage sexual pleasure. In short, Carrie learned that God wants her to like sex, to "just have fun in the marriage bed." Carrie credited this transformation to both LCL and her own spiritual devotion: "I would say it was 30 percent LCL and 70 percent doing [spiritual] battle and praying."

CarrieForChrist called her story a sexual awakening. Sexual awakening stories are well established in the vernacular of Christian sexuality websites. Like evangelical salvation narratives or testimonies, they follow a distinct formula: the narrator lives through a time of sin and suffering that he or she then overcomes by believing in God, who has the power to transform believers' sexual lives. LustyChristianLadies.com has even provided its readers an instructional blog post on the topic, "How to Have a Sexual Awakening." The post describes the experience as "a sudden revelation of God's intention to have a richer sexual relationship with [one's] husband." Blogger Kitty describes the early years of her marriage, when she had only a "minor interest in sex" and didn't communicate about it with her husband. Then, "quite all of a sudden and surprisingly," she experienced a sexual awakening. She credits God with her transformation, and tells her readers that faithfulness is key to achieving sexual fulfillment: "The most practical thing you can do to change is to pray continually for God to change you. He is on your side. He wants your spouse to be free even more than you do. Ask Him to make you who you need to be in order to be a blessing to your spouse. Do all that He leads you to do." Although she places change and transformation ultimately in the hands of a divine creator, Kitty also tells her readers to actively pray and urges them to *do* all that God leads them to do. Sexual awakening stories, like salvation stories, deftly combine a sense of human agency with submission to God's will. As Virginia Brereton argues about salvation narratives, conversion requires an *actor*, someone who "accepts Christ" rather than "is accepted by Christ." This centralizes the responsibility of individuals when it comes to their own eternal fate.[1]

How believers imagine themselves as actors, rather than acted upon, depends on how they tell their religious stories. In this chapter, I analyze how some Christian women interpret their sexual experiences by describing them according to a particular narrative form. Like creators of Christian sexuality websites, who emphasize how their actions align with their faith to justify the sexual content on their sites, women tell sexual awakening stories that align their sexuality with their evangelical Protestant beliefs. They make their unique experiences conform to the particular narrative components of obstacles and redemption that make up the before and after of the awakening experience. This points to the importance of personal piety, the marriage relationship, and Christian sexuality websites themselves in shaping what is sexually possible and permissible in a Christian setting. In telling sexual awakening stories, women prioritize their choices and desires, although they do so in a way that fits an evangelical mold.[2]

Though both men and women tell stories that they call sexual awakenings, these narratives are uniquely positioned to give voice to women's experiences. I do not analyze men's stories in this chapter for two reasons. First, the vast majority of sexual awakening stories are told by women, and I have only limited data on men's stories. Men make references to their "awakenings," but there are few detailed narratives.[3] Second, and more important than the *quantitative* differences in the number of stories told by men versus by women, men's stories are qualitatively different than women's. Despite gender-equal language that permeates the logic of godly sex, men and women who use Christian sexuality websites present their stories on different and imbalanced trajectories. Secular and religious talk about sexuality recognizes men as sexual and encourages men's heterosexual desire for (and access to) women. Christian men are not removed from their sexual identities in the same way as Christian women, making it more difficult for men to tell stories that contain the narrative components important to a sexual awakening story. In other words, men are already sexually "awake" when they become sexually active within marriage.

Women's stories suggest that women's bodies and the pleasure they experience are deeply connected to others—God and their husbands—and that they must balance their own needs with selfless acts that prioritize their marital relationships and family. This maintains gender imbalances between men and women and restricts women's sexual expressions. Contradictory messages of sexual entitlement and selflessness within women's sexual awakening stories serve to situate them within a conservative Christian culture that continues to perpetuate gender hegemony. Reflecting a postfeminist sentiment that combines anti- and pro-feminist messages, Christian sexuality websites are places where women make sense of sexual pleasure in multiple ways without challenging male privilege within their sexual relationships. Sexual awakening stories show how women both theologize and sexualize their bodies to make sense of the pleasure they believe should be a part of Christian marital intimacy.[4] Their stories are as much about the relationship between the body and religion as they are about the body and sex.

Women's Pleasure

In contemporary America, women's sexuality shows up in all kinds of unlikely places. It appears in expected red-light spaces—through pornography, erotic dancing, and sex work—but also in spaces that are quite ordinary, even "wholesome." There are at-home sex toy parties organized by suburban housewives; fitness centers that offer pole dancing exercise classes; and vibrators sold at chain pharmacies like Walgreens. Talk of empowerment often exists alongside these depictions of women's sexuality. Popular media depicts secular, white women as in control of their sexuality and free from gender inequality. Feminism—at least the kind that equates sexual autonomy and pleasure with women's freedom—has gone mainstream.[5]

Women's entitlement to sexual pleasure was central to second-wave feminism; if bad sex (forced or obligatory) signaled women's oppression, good sex on women's terms was a part of their liberation.[6] Yet contemporary representations of women's sexual pleasure have largely lost their political and radical edge. This is indicative of what some scholars call *postfeminism*, a cultural trend that merges anti- and pro-feminist ideas that give women a sense that they control their sexuality while at the same time encouraging a sexuality that acquiesces to men's interests. Women who boast sexual confidence do so within a social structure that permits ongoing sexual violence and maintains gender imbalances in education, at the workplace, and at home.[7] Despite what often appears to be gender-equal language, popular discourse supports and expects gender difference that tends to privilege men, especially when it comes to sexual desires and expressions.

When this "common cultural script" meets evangelical Christianity, it becomes, in the words of sociologist Michelle Wolkomir, a "divine mandate."[8] Christian sexuality website users construct a godly sexuality for women akin to what Rosalind Gill calls "compulsory (sexual) agency"—the contradictory notion that women feel social pressure to *choose* to improve their sex lives.[9] Although these users emphasize the mutuality of sexual pleasure (see chapter one), for Christian women, being "sexually awakened" means experiencing pleasure within a very specific, male-dominated context. Nonetheless, Christian sex advice uses religious beliefs to justify women's pleasure. Authors Ed and Gaye Wheat, for example, write that the ability to orgasm is what "God designed for every wife." Shannon Ethridge tells women that "sexual confidence isn't just for the supermodel or porn star. It is the birthright of every woman." In fact, Ethridge would say that sexual confidence, as envisioned by God, is *not* for supermodels and porn stars at all but *only* for Christian wives.[10]

Evangelicals write about women's pleasure—describing it as "mysterious," "elusive," and "just out of reach"—to demystify it. Christian sexuality websites and sex advice books offer women and their husbands the tools to help women achieve physical pleasure: step-by-step instructions on how to arouse a woman, anatomical drawings identifying the clitoris, advice on lubricants, suggestions about what time of day to have sex, lists of romantic gestures, and descriptions of sexual positioning—all intended to optimize women's pleasure. Just as authors did during the feminist movement of the 1970s, Ethridge, in *The Sexually Confident Wife*, writes candidly about clitoral orgasms. She tells women to "delightfully indulge in the pleasure of the moment" and instructs wives to allow their husbands to focus on making them aroused before having sexual intercourse: "Let him manually, visually, and orally explore your private playground, showing him how you'd like to be touched if necessary. Don't feel rushed to reciprocate yet. Just enjoy the pleasure signals your body is sending your brain right now. Let this pleasure nourish your spirit and draw the two of you closer emotionally." Ethridge prioritizes women's bodies and pleasure within the sexual relationship. She gives them permission to be selfish—even if just for a moment. Yet unlike women's liberationists, Ethridge carefully contextualizes pleasure as being good for women's spiritual and marital lives, making both God and women's husbands key to women's experiences.[11]

Women's stories discuss sexual pleasure in ways that parallel a feminist sensibility about women's entitlement to pleasure and their bodies while reflecting a conservative Christian sensibility about the role of marriage and God in women's lives. Ethridge writes positively about female pleasure, even going so far as to suggest women's natural potential for pleasure exceeds that of men. *The Sexually Confident Wife* includes information like, "Did you know the female clitoris has eight thousand nerve fibers? That's almost twice as many as the male penis!" Ethridge quotes secular science writer, Natalie Angier, who writes, "[Some women] never bought Freud's idea of penis envy; who would want a shotgun when you can have a semiautomatic?" Women's sex organs—the semiautomatics—hold the potential for intense and long-lasting pleasure. Yet at the same time, Ethridge frames what she describes as exceptional female pleasure potential as only possible within the pleasure of the marriage relationship:

Women have the luxury of a much shorter refractory period, which means she can be an orgasmic Energizer bunny and keep going and going if she wants to. A woman's body is capable of experiencing these intense waves of pleasure over and over for several minutes [. . .]. Usually, it's an overwhelming desire for intercourse with her husband that brings these orgasmic waves to an end, as she demands he replaces his fingers with his penis.

In explaining G-spot orgasms and the potential for multiple orgasms, Ethridge first focuses only on women's bodies and the pleasure women can experience. Ultimately, though, she describes a woman's pleasure—however powerful and long lasting—as inevitably leading to an equally intense desire to be penetrated by her husband. Ethridge gives women agency in this scenario—a woman "demands" that her husband penetrate her with his penis—but limits women's choices to this quintessential act of male sexual dominance. As she states clearly in the subtitle of the book, Ethridge defines sexual confidence as "connecting with your husband—mind, body, heart, spirit."[12]

Sheet Music author Kevin Leman writes extensively about women's orgasms but also prioritizes women's pleasure vis-à-vis men's. In the chapter "The Big 'O,'" he writes admiringly about women's bodies and the pleasure they experience: "Many women are surprised when I tell them that a large percentage of men are jealous of their orgasms." He goes on to describe women's orgasms magnanimously: a woman having an orgasm feels like "the world is exploding" and she is "riding the waves of ecstasy." Yet he describes women's pleasure as ultimately benefiting the self-image of men:

> Women, this might surprise you, but even more than your husband wants to have sex with you for his own sexual relief, the truth is, he wants to please you even more than he wants to be pleasured. It might seem like it's all about him, but what he really wants, emotionally, is to see how much you enjoy the pleasure he can give you. If he fails to do that, for any reason, he'll end up feeling inadequate, lonely, unloved.

Leman frames women's pleasure as a way for men to prove their sexual prowess—to show "the pleasure he can give you." Although he prioritizes women's pleasure within the marriage relationship, it is not for women themselves but rather for the benefit of men, so they do not feel "inadequate, lonely, unloved." Leman's repeated comment that he might "surprise" women with his information suggests that they do not already know much about their bodies.[13] Instead, Christian women need male experts to inform them.

As much attention as popular Christian authors give women and their orgasms, women appear to have trouble applying this prescriptive advice to their lives. Women who use Christian sexuality websites often join these sites because they suspect they should enjoy sex but don't know how. Stories of sexual awakening trace the process by which this cognitive knowledge about God's design for sexuality becomes *embodied* knowledge. As one woman who shared her sexual awakening story on BetweenTheSheets.com described, "I knew when I got married that sex wasn't dirty or sinful. At least I knew this in my head, but it just never worked its way through my subconscious." Sexual awakening stories explain how the body transforms to reflect what these website users already believe in their minds. Whereas prescriptive Christian sex advice gives women permission and guidelines to experience pleasure, online discussions go further to help women to overcome their unique obstacles and circumstances.

The Imperfect Body: Before the Awakening

Because sexual awakening stories are always told after women have experienced an awakening, hindsight allows women to make meaning of the obstacles that prevented them from experiencing sexual pleasure. Whether these obstacles are the result of past sexual sins or physical ailments, sexual awakening stories consistently present women's bodies as their source. In chapter one, I described what I call an inhibition paradox, which simultaneously encourages and condemns Christians' sexual pleasure. This is especially true for women, who hear a constant refrain of messages that downplay or vilify their sexuality. Sexual awakening stories show how women *inhabit* the inhibition paradox. They internalize and individualize it, describing distinct physical, emotional, and spiritual barriers to their sexual pleasure. The body—which is the catalyst for sexual pleasure and marital wholeness—is also the barrier that prevents women from achieving sexual pleasure.

Even though conservative Christian messages condemn sexual activity outside of marriage unequivocally, both for men and women, these messages frame men's sexual desires as natural and expected but are relatively silent when it comes to women having desires of their own. This compounds the inhibition paradox for women; they may experience sexual desire but feel guilty or self-conscious about it, even in the "proper" confines of marriage. Samantha, owner of the online sex-toy store, describes this pointedly:

When sex is talked about in church, it's talked about like this: men have sexual needs and women have emotional needs. And nobody talks about the fact that someone with ovaries may indeed have a sexual need EVER. And I want to raise my hand and go, 'excuse me!' It's just so not talked about. And if it's only talked about from the pulpit that men only have sexual needs, then that means that women's needs (a) don't exist or (b) aren't important to God.

Christian men are not removed from their sexual identities in the same way that Christian women are. Even men who have never engaged in sexual acts, Samantha points out, are more likely to have been exposed to positive sexual talk geared toward them. Sexual awakening stories reveal how men and women set out on different and uneven sexual trajectories.

Christian women do not receive positive messages about their sexuality from church, and they don't receive it from secular culture, either. Evangelical women who are "in the world" but not "of the world" must make sense of secular messages that they are exposed to but that shouldn't apply to them. One LustyChristianLadies.com reader, XYZ, called this the world's "worship of sex," explaining, "For much of the unsaved world, sex has become a 'God.' They worship the creation of sex rather than the *creator* of sex." Many women website users are particularly critical of secular depictions of women's sexuality, calling them ungodly. Blogger Maribel told me that she created her blog, MaribelsMarriage.com, because she believes that secular messages that sexualize women inadvertently make Christian women feel like they shouldn't be sexual: "I think a lot of Christian women have a lot of guilt with sex. It's often referred to as the 'good girl syndrome,' where they don't think they're a good girl if they're enjoying sex because they've been told their whole life 'no, no, no, no you shouldn't be doing this. Good girls don't have sex.'" What Maribel describes as "good girl syndrome" adds a gendered critique to the inhibition paradox: women's unique inability or hesitance to enjoy sex in marriage.

Before experiencing a sexual awakening, Christian women describe many contrary sources of inhibitions. A religious upbringing may lead women who try to experience sexual pleasure in marriage to feel guilt, insecurity, and a lack of knowledge, but an upbringing without religion can skew women's sense of their own sexuality and what is godly. A past of sexual sins can get in the way of a woman's current sexual relationship just as much as a past of abstinence may prevent a woman from optimizing her sexual pleasure by stunting her as a "good girl." These inhibitions affect who women are and who they think they should be. Tara, a LCL reader, put it this way: "Christian women know they don't want to be Carrie Bradshaw [the promiscuous New Yorker from the hit TV show *Sex and the City*], but they don't want to be prudes either." Finding space in between—to be sexual in the way that God approves—is difficult for women who experience disconnect between their religious beliefs and sexual desires.

Notes

1. Brereton, *Sin to Salvation*, 48.
2. There is a rich body of research investigating how conservative religious women exhibit agency within the constraints of their patriarchal religions. For a review of this literature, see Kelsy Burke, "Women's Agency in Gender-Traditional Religions: A Review of Four Approaches," *Sociology Compass* 6 (2012): 122–133. For an evangelical context, see Brenda E. Brasher, *Godly Women* (New Brunswick, NJ: Rutgers University Press, 1997); and R. Marie Griffith, *God's Daughters: Evangelical Women and the Power of Submission* (Berkeley, CA: University of California Press, 1997).
3. In a BTS thread asking members to share the stories of their sexual awakenings, over a hundred members posted their stories, and all but one were written by or about women.
4. For more on the importance of sexual stories, see Kenneth Plummer, *Telling Sexual Stories: Power, Change, and Social Worlds* (New York: Routledge, 1995). Jill Peterfeso observes how important these stories are for making sense of religious women's sexuality in "From Testimony to Seximony, from Script to Scripture: Revealing Mormon Women's Sexuality Through the Mormon Vagina Monologues," *Journal of Feminist Studies in Religion* 27 (2011): 31–49. In the article, Peterfeso describes how performers draw from the rhetorical form of *The Vagina Monologues* to critique the Mormon Church, celebrate women's sexuality, and still remain devout believers of their Mormon faith.
5. Journalist Ariel Levy argues that the sexualization of women through what she calls "raunch culture" is anti-feminist. According to Levy, the goal of women's liberation—to empower women sexually—never actualized: "The truth is that the new conception of raunch culture as a path to liberation rather than oppression is a convenient (and lucrative) fantasy with nothing to back it up" (Female Chauvinist Pigs, 82). A more optimistic reading of the intersection of feminism and pop culture can be found in J. Jack Halberstam's *Gaga Feminism: Sex, Gender, and the End of Normal* (Boston, MA: Beacon Press, 2012).
6. In *Desiring Revolution: Second-Wave Feminism and the Rewriting of American Sexual Thought, 1920 to 1982* (New York: Columbia University Press, 2013), Jane Gerhard outlines the history

of second-wave feminism and women's sexual pleasure, arguing that "feminists agreed on little else beyond the shared value of women determining for themselves what they wanted from sex" (8).

7. Burkett and Hamilton, "Postfeminist Sexual Agency"; Gill, "Empowerment/Sexism"; Levy, *Female Chauvinist Pigs*.
8. Michelle Wolkomir, "Giving It Up to God: Negotiating Femininity in Support Groups for Wives of Ex-Gay Christian Men," *Gender & Society* 18 (2004): 739.
9. Gill, "Mediated Intimacy," 33.
10. Wheat and Wheat, *Intended for Pleasure*, 111; Ethridge, *Sexually Confident Wife*, 13.
11. Ethridge, *Sexually Confident Wife*, 106.
12. Ibid, 109, 113, 112.
13. Leman, *Sheet Music*, 91–92, 11.
14. Driscoll and Driscoll, *Real Marriage*, 124.
15. Ibid.
16. DeRogatis, *Saving Sex*, 54.
17. Ethridge, *Sexually Confident Wife*, 108.
18. In Born *Again Bodies*, Griffith argues that the body is central to twentieth-century Christianity. Control of the body (especially in regards to sexuality and diet) has exposed "the complex relationship between the visible body and the invisible soul" (23).
19. Attwood, "Sexed Up," 87.
20. Anthropologist Saba Mahmood's ethnography of women participating in the Egyptian Islamic mosque movement, *The Politics of Piety: The Islamic Revival and the Feminist Subject* (Princeton, NJ: Princeton University Press, 2005), is instructive when considering religious women's ability to make choices when faced with constraints. She writes that religious women's accounts must be analyzed according to "the particular field of arguments" made available by their religious communities and "the possibilities for action these arguments have opened and foreclosed" (183). Mahmood diverges from typical definitions of agency as being defined by free will and instead considers an agency that may be docile and compliant, reflecting the possible choices available to women given their religious circumstances.
21. Stevi Jackson and Sue Scott describe the "sexual sentence," a typical narrative about sexual activity that focuses on vaginal intercourse between a man and woman, leading to mutual orgasm and ending ultimately with a man's ejaculation. "Faking Like a Woman? Towards an Interpretive Theorization of Sexual Pleasure," *Body and Society* 13 (2007): 95–116.

DISCUSSION QUESTIONS

1. Describe the upbringing of the user called CarrieForChrist. What did she discover online?
2. What did CarrieForChrist learn online about other Christian women? What did the website allow her to realize?
3. According to the article, do women feel that the Evangelical websites allow them to explore their sexuality without "challenging the male privilege" in their sexual relationships? Do you agree?
4. Describe what can be considered a "Godly sexuality."
5. Explain what information is found within *The Sexually Confident Wife*.
6. Describe the idea that women in Christianity might feel guilt over sexual desire. Do you agree?

ISLAM

INTRODUCTION

Hajj is an "annual pilgrimage to Mecca," and as one of the five pillars of Islam it is a journey that all Muslims should take at least once in their lives. This pilgrimage is a transformative trial of the soul, in which Muslims ultimately seek forgiveness from God. They psychologically process their sins, their lives, and their deaths—always alongside a united Muslim ***umma,*** or "community." According to Muslim tradition, during *hajj* season millions of Muslims perform rituals reliving the story of Abraham, his wife Hagar, and their son Ishmael. Muslim heritage also comes from a long line of prophets, beginning with the prophets in the Hebrew Bible and ending with the **Prophet Mohammed** PBUH ("Peace Be Upon Him"), who is known as the **seal of the prophets**. For Muslims, the Prophet Mohammed PBUH would have been the most perfect Muslim who ever lived. Knowing and emulating how the Prophet PBUH performed the *hajj*, among most other forms of Muslim practice, is considered a priority for believers.

Muslims desire to follow in the footsteps of the Prophet PBUH, and they see him as a restorer of Abraham's practice and worship of one God. Stories about what the Prophet PBUH did, said, and silently approved of have been authenticated and canonized and are known as ***hadith***. They serve as a secondary sacred source. The Prophet PBUH was born in 570 CE. He grew up in Mecca, a city that arose through commerce and trade. As money flowed into this mercantile city, much of the wealth and power favored particular tribes, resulting in an unjust system being overseen by a small oligarchy. These changes and the consequent distress they caused for most people caused the Prophet PBUH to meditate in and around the caves of Mecca. One night in 610 CE, at **Mount Hira**, he received the first of many revelations. This night, known as the **Night of Power and Excellence**, is commemorated during Ramadan. The Prophet PBUH continued to receive revelations until his death in 632 CE.

Ramadan is a "holy month during which Muslims believe the devil is chained away." During this month, Muslims spend their days fasting, another pillar of the Islamic faith, and also abstaining from sex, cursing, smoking, and other physical desires. Spiritual pursuits take center stage as Muslims pray, recite, and meditate on the Qur'an more than they normally do. The **Qur'an** is the "primary sacred text within Islam" and is a compilation of the revelations received by the Prophet PBUH. The Qur'an was compiled decades after the Prophet's PBUH death. It is arranged with no chronological order; rather, it is collected in 114 *surahs*, or "chapters," ordered according to length, from longest to shortest. For Muslims, every revelation comes directly from God, and the **Angel Gabriel** served only as a vehicle to facilitate that transmission. Further, for Muslims the Qur'an is the Qur'an only when it is in the Arabic language; a translation is not considered to be the Qur'an. By the same token, the Arabic from the Qur'an is so elevated, rhythmic, and complex that it eventually came to be used for grammatical instruction.

Many Muslims also believe that the Prophet PBUH was illiterate, and they see this masterful work the Qur'an as evidence of God's true authorship—a true miracle. In Ancient Arabia many of these revelations were revolutionary and caused controversy and anger, specifically the revelation that there is only one God. The concept of monotheism proved to be a difficult revelation to spread throughout Ancient Arabia, where nomadic tribes and their beliefs in animism and multiple gods were commonplace. For Muslims, this period of time is known as *jahiliyyah*, or "the time of ignorance." This revelation of monotheism was also especially troublesome because the pilgrimage to Mecca was being made before Islam spread in Ancient Arabia. During the lifetime of the Prophet PBUH, however, the *Ka'bah*, located at the center of Mecca, was the home of more than 300 gods. The pilgrimages to this cite included buying animals in sacrifice, and for the nomadic tribesmen who were the keepers of the *Ka'bah*, the idea that there was only one God also posed an economic threat.

Today, the ***Ka'bah*** is "a symbol of the worship of the one God during *hajj*." A large cube-shaped building, for much of the year it is covered with an enormous, heavy cloth. Notably, in one of its inside corners is a revered black stone. The *Ka'bah* orients all Muslims for prayer, no matter where they are in the world. For many Muslims, Adam built the *Ka'bah* and Abraham rebuilt it with his son Ishmael. Muslim tradition relates the heavy persecution experienced by the early companions of the Prophet PBUH in Ancient Arabia as both an ideological threat and an economic one. Some of the the Prophet's PBUH companions felt the persecution so strongly that they fled to Ethiopia. There was even an attempt to murder the Prophet PBUH. Yet in the middle of all of this, the city of Yathrib, today known as **Medina**, heard of the Prophet's PBUH reputation. The tribes at Yathrib needed an arbiter to judge internal conflicts, and they called upon the Prophet PBUH to become that judge. So in 622 CE, the Prophet PBUH and many of his companions left Mecca for Yathrib.

This "great migration" is known as the *hijrah*. The *hijrah* is a seminal moment in Muslim history, as it marks their calendar; that is, there are years before *hijrah* and years after. *Hijrah* is important because it is in the city now known as Medina that the first Islamic community was born. This speaks to the *umma*'s sacred role within this religion. Yet as the new Islamic community grew in Yathrib, the Meccans did not stand down. There were a series of battles between the new Islamic community and the Meccans, eventually leading to a truce. When the Meccans broke the truce, in 630 CE, the new Islamic community was strong enough in numbers to storm the city of Mecca. The Prophet's PBUH first act within Mecca was to clear the *Ka'bah* of idols and restore Abraham's initial intention of worshipping one God. Today the Prophet's PBUH burial ground is located in Medina, and many pilgrims visit this site before landing in Mecca for *hajj*.

After the Prophet's PBUH death, many of the tribes of Ancient Arabia were under the Islamic banner. But the temptation was spreading to no longer validate the ties made with the fallen Prophet PBUH. Now the issue turned to who would lead the community and keep its people unified. **Sunni** Islam is the largest Muslim denomination, at almost 80% of the whole community. The early Sunni understanding of proper leadership came about during the time of the four Rightly Guided Caliphs. A **caliph** is a "spiritual leader of Islam, claiming succession from Mohammed." In addition to leading in spiritual matters, caliphs were expected to function as statesmen, diplomats, and generals and exercise fiscal control. Under the two earliest rightly guided caliphs—**Abu Bakr** and **Umar**—Islamic expansion progressed quickly. In the last years of the reign of the third rightly guided caliph, **Uthman**, dissension grew within the Islamic community. Uthman was murdered, and some feared that the fourth rightly guided caliph—who was named **Ali** and was the cousin and son-in-law of the Prophet PBUH—would not bring the murderers to true justice.

A rift occurred that eventually led to civil war. Those who sided with Ali became known as the party of Ali, and they moved the capital from Medina to Kufa. It is from this moment that **Shi'a** Islam, the second denomination within Islam, began to grow. During a power grab, Ali himself was murdered, and the time of the Rightly Guided Caliphs ended. The capital was moved to Damascus, and the reign of the **Umayyad Dynasty** began. The expansion continued under Umayyad leadership, and it even reached Spain. The Umayyads also centralized and "Islamized" their empire. This type of leadership continued into the **Abbasid Dynasty,** whose leaders moved the capital to Baghdad. The Abbasids set up translation centers that preserved much of the Greek philosophy that would later play such a huge role in the European Renaissance. Under the Abbasids, intellectual growth increased in every field of study, especially at learning centers and mosques. Although Europe was in the Middle Ages, in Islamic history this period of time was marked with the light of growth, innovation, science, and medicine.

Much has happened since then, leading to a huge spectrum within both denominations of Islam. Different *hadith* for Sunnis and Shi'as have led to a huge difference in practice and application. Some Shi'as developed an elaborate clerical system, whereas clergy is mostly absent in Sunni Islam. In some forms of Shi'a Islam, Muslims pray to intercessors and have images depicting them, while in Sunni Islam these practices are considered blasphemous. However, even within Sunni Islam itself, there are several legal schools of thought and several movements, some more extremist than others. Shi'a Islam also lives in different forms, all with specific practices that differ from one another. As the world's fastest-growing religion, Islam is facing several challenges in the 21st century. Despite these differences and challenges, many Muslims are calling for a return to the essence of Islam. Islam is a religion whose name means "submission and peace"—a submission to God that leads to a peaceful way of life that should fill believers with a resolve to love one another.

ECOLOGY

The fossil fuel industry is a major perpetrator of the current climate change disasters. The Muslim Middle East is at the front of the oil sector within this industry. As billions of people around the world depend on oil, the environmental crisis deepens. Although Muslims might not be associated with ecological concerns, there are many Muslims, like Ibrahim Abdul-Matin, who claim that their faith and sacred text show Islam as being innately "Green." For them, Islam is interested in protecting the environment, and Abdul-Matin has written a book about the different ways Islam calls Muslims to environmental action. The following excerpted chapter looks at the concept of "Green Muslims."

ARTICLE 1

GREEN MUSLIMS

By Ibrahim Abdul-Matin

I was inspired to write this book after reflecting on my own Green Deen and meeting other Green Muslims who are living the six principles of a Green Deen. I sought out Muslims who are committed to being stewards of the Earth (*khalifah*), who understand the Oneness of God and His creation (*tawhid*), who look for signs of Allah (*ayat*) in everything around them, who move toward justice (*adl*), who seek to protect the delicate balance of the natural world (*mizan*), and who honor our sacred trust with God to protect the planet (*amana*). Happily, what I discovered is that Muslims are involved in every aspect of the stewardship of the Earth.

Stewardship of the Earth comes in many forms. Green Muslims like Aziz Siddiqi of Houston, Texas, are actively involved in environmental policy. Others, like Sarah Sayeed of the Bronx, New York City, are Green Muslims who ensure environmental justice by working with the interfaith environmental community. While this whole book reflects the active involvement of Muslims in the environmental movement, this chapter focuses on some distinct and inspirational efforts, including the famous DC Green Muslims in our nation's capital.

The Color Green

Muslims have a personal connection to the color green. Color is a refraction of light. In Islam, light is the substance of creation. Somewhere, in the farthest reaches of the universe, Allah is creating from pure light. Green is an aspect of that light and is reflected all over the world. The favorite color of the Prophet Muhammad (peace be upon him) was green: "Among the colours, green was liked the most, as it is the colour of the clothing in Jannah (paradise)."[1]

Allah paints his *ayats* (signs of nature) in a tapestry of green all over the world. He does so from the lushness of Muir Woods in northern California to the evergreens and ferns of the East as they welcome the flowers after spring rains. Green is the color of the Waipio Valley in the heart of the Big Island, Hawaii. In this valley, you can dip, drink, and make ablution (*wudu*) in freshwater streams surrounded by green. Green are the mountainous ridges of the Rockies, covered with ferns that host the snow in winter. Green is the color of life on all corners of the planet. And, yes, people can be green too.

Are You a Green Muslim?

Your own family might be the best place to look for signs of a Green Deen in action. That's certainly where I look first. I come from a family in which our father taught us that the Earth is a mosque. My father spends much time in the out-doors being absorbed by the natural world. Connecting to nature is how my father connects with God.

One morning while writing this book, I received the following e-mail from my dad. He had just arrived at a remote section of Maine for one of his customary sojourns into the natural world. His goal was to be immersed in the signs of Allah and to pray and worship the Creator while surrounded by those signs, free from distraction.

> Alhamdullilah [All Praise Due to Allah], I arrived in Maine safely. Tonite I am spending the night at Sebago Lake (car camping). Tomorrow I will be traveling north along the coast to the ocean. Not sure how many nights I will spend there. Will figure that out when I get there. I am expecting Wednesday to be a rainy day, but although it's partially cloudy, the weather is nice. Sebago Lake is a huge lake—I even heard they have a navy SEALs training base somewhere near here. I probably won't get a chance to e-mail after today, so I'll check in again Thursday, Insha'Allah [God Willing]. Make Dua [supplication] for me, I will do the same for you.

Daddy (yes, I still call him that) sent that message to all six of us children. He is the best man I have ever known. Of course I know he is not perfect, but he is still my father. He has always done his best to guide and protect me. I cannot say that I have never disappointed him, but I have never raised my voice to him. I am his son through and through, and I love what he loves as much as I love him. Trying to be a steward of the Earth is one way I am following in his footsteps.

We are a proud people rooted in the land. My father's family comes from Virginia, where my grandmother was one of nine in a traditionally well-educated churchgoing black family. Back in Scottsburg, Virginia, you can still see the grave markers from my extended family going back to the 1860s. My father's father, my grandfather, was born on a Native American reservation in upstate New York in 1908.

My father is a man of firm principle. He chose the path of Islam and made his life's motto "There is no God but Allah and Muhammad is His Messenger." This is the declaration of faith in Islam, the one thing a person must say and believe to be considered Muslim. Like my dad, I am a Muslim. As a child of converts to Islam, I did wonder if my parents' Islamic path was the right path for me. I spent a long time pondering this question. I became aware of other systems of thought, belief, and practice. I believed that humans could not have created and developed this world alone. I was not an atheist—I believed in an omnipotent Creator. Eventually I decided that Islam was the best method for decision making I could find, that it was what I wanted to guide me in my life and the religion I wanted to raise my children in one day.

My dedication to the environment, my commitment to being green, starts with my father and stays alive by my Islam. I try my best to not separate what I do and how I live from what I eat. I try my best to treat the planet as sacred, like the mosque I believe it is. Finally, I try to better my Green Deen by learning from others who choose to root their love of the planet in their faith.

The Pioneer

Aziz Siddiqi is a Green Muslim who has been on the forefront of environmental policy since the 1970s.[2] I will refer to Mr. Siddiqi here as Uncle Aziz, "Uncle" being a term of respect in many cultures. He is a pioneer of sustainability who has been working to ensure that the air you breathe is properly protected and regulated. It was his orientation as a Muslim that gave him the perspective to focus on the basic principles of balance (*mizan*) and Oneness (*tawhid*), which helped him become one of the most important figures in the history of the environmental movement.

In the late 1960s, Uncle Aziz was a young doctoral candidate doing groundbreaking research in chemical engineering. His work was noticed by an official from the University of Houston, and a few years later, upon completing his studies, he was immediately offered a job there. Soon he found himself guiding the development of a curriculum that would help the U.S. Environmental Protection Agency (EPA) carry out its new mission of enforcing the Clean Air Act.

In 1973 the EPA was only three years old and did not understand the full breadth of its power. The EPA, its scientists, and its partner agencies needed to be trained on how to monitor pollution from smokestacks and other commonly used industrial practices. Uncle Aziz had to learn how to explain his research in chemical engineering to this group of regulators. He also authored the training materials used to teach EPA scientists how to sample ambient air and develop pollution controls.

Eventually, Uncle Aziz became a national authority and a local community leader. He lectured widely and wrote articles that became canon in the industry. His status as a well-respected scientist working to protect the planet was balanced with a reputation for being a pious member of the Muslim community of Houston. In time he used his contacts, expertise, passion, and background to start his own consulting firm, which he runs today.

Along with being credited with helping Houston significantly reduce its smog levels over the years, Uncle Aziz is the president and CEO of one of the largest Islamic institutions in the United States—the Islamic Society of Greater Houston. Under his leadership, the society is using its multimillion-dollar budget to buy land and develop mosques that include a community center, a school, and a free clinic for the general public.

Uncle Aziz embodies the spirit of protecting the planet, protecting the Earth, and praising Allah. His mind never rests, for he is constantly innovating. Today, he is thinking about starting a new venture concerning energy efficiency and conservation.

Interfaith Green Muslims

Interfaith work is one of the best places to exercise a Green Deen. Most faith traditions believe that humans will be held accountable for their actions. Individual responsibility reflects the trust (*amana*) we entered into with God when we were blessed with the gift of choice.

Allah says in the Qur'an:

> That Day will Man be told (all) that he put forward, and all that he put back. (Qur'an 75:13)

"That Day" refers to the Day of Resurrection, and the rest points to humanity's actions while on Earth. The notion of a life after death is a strong organizing principle for Muslims, Buddhists, Hindus, Christians, and Jews. Many people of faith are mindful of their actions as a way to ensure a peaceful and heavenly afterlife. It has been said that faith can speak the common language in the public square—a belief that speaks to the ability of people of faith to create something that everyone can be a part of.

Sarah Sayeed is a Green Muslim who lives in the Bronx, New York City.[3] Her neighborhood has seen the best and worst of times in New York City's recent history, and Sarah is most committed to the Green Deen principle of justice (*adl*). Following this principle, she helps to organize people of all faiths for environmental justice.

Sarah works at the New York Interfaith Center, located on the Upper West Side of Manhattan adjacent to the Columbia University campus. The center's unique nickname, "the God Box," was coined when people of faith began moving their organizations into the building. As a Muslim activist working from the God Box, Sarah is connected to just about every relevant Muslim leader in New York City. She understands the nuance of secular and religious culture, is well educated, and loves New York. To Sarah, New York City is a mosque.

Sarah is one of the core organizers of the Faith Leaders for Environmental Justice, a gathering of faith leaders—Buddhist, Jewish, Christian, and Muslim—who are deeply concerned about the environment, particularly climate change and food security. Over the past three years, these leaders have come together on a regular basis to cultivate a common language of stewardship (*khalifah*). According to Sarah, stewardship of the Earth can compel people of faith to engage in interfaith work without having to agree on basic "creed."

The Faith Leaders share information on pressing issues of food scarcity, solid waste management, energy use, and green jobs. On Earth Day 2010, they launched their "green map," which identifies locations throughout Harlem offering access to healthy food to people trapped in New York City food deserts—areas without access to fresh foods.

Interfaith work can offer a clear message to policy makers about the moral imperative. "Politicians," says Sarah, "need to hear the moral reasoning to remind people of the moral need to act in a just way beyond what they would normally do." Lisa Sharon Harper, founder of New York Faith and Justice and one of Sarah's co-conveners says, "The faith voice gives gravity to the numbers."[4] The lesson to policy makers is the same—overcoming any challenge requires coalition building.

Sarah once told me that all spiritual paths tread upon the Earth. I had the honor of watching many of these paths converge at an interfaith Ramadan event in 2009, at which I moderated a panel on pollution with a Jewish scholar, a Christian educator, a Jain activist, and a Muslim imam.[5] Muhammad Hatim of New Jersey used the words *mischief* and *corruption* synonymously with *pollution*. "Are humans here to pollute?" he asked. "Are they here to cause corruption and mischief? Are we responsible for the state that the planet is in ecologically? What are the signs that can tell us how we are to live on our only planet?"[6] Those questions struck me primarily because I realized that each faith had faith-specific answers to those questions. Each panelist went on to describe what they felt were their responsibilities to the planet. I reflected on Islam and recalled the verse of the Qur'an that says:

> *lakum deen oo kum waliyy ud deen*: "to you then be your way and to me mine." (Qur'an 109:6)

All of the Deens—the different spiritual paths—that day at the interfaith event recognized their coexistence on the planet. They each made an unwavering commitment to restoring justice to the Earth.

The DC Green Muslims

The Washington, DC, Metro area has become a hub of Green Deen activity.[7] The ADAMS (All Dulles Area Muslim Society) Center Mosque is home to the DC Green Muslims, a loosely organized group of like-minded Muslims interested in stopping the detrimental effects of climate change. They exemplify the principles of justice (*adl*) and trust (*amana*) and are committed to both personal adaptation and collective responsibility. They are a politically active community and truly see the DC Metro area as a mosque. The center has also just published its *ADAMS Center Green Environment Guide*.[8]

The DC Green Muslims foster fellowship, and my first interaction with them was at one of their famous dinners, held in a building that was once an opulent home but which is now owned by the DC Parks Department. This particular dinner was a networking event for young professionals—a way for Muslims to connect to the larger green movement, to be engaged in the political process, and to be involved in something that affirms and strengthens their Deen. My sister Tauhirah, who works on water quality issues, and I drove down together from New York to attend the dinner. (I'll be discussing Tauhirah's work further in part III of this book.) I learned that the group's first Ramadan Iftar—the breaking of the Ramadan fast—consisted of only fifteen people. Their first conversations were on a broad scale, dealing with larger concepts and the intersections between Islam and the environment. That small group had grown to the event I attended, which hosted more than two hundred people, and the discussions there focused on tangible steps these young leaders could take to activate their broader community to build a Green Muslim movement.

The DC Green Muslims have participated in local greening efforts alongside non-Muslim environmental organizations. For example, they collaborated with the DC Parks and People to plant trees and clean up a local park. Sarah Jawaid, a DC Green Muslim and a key DC Green Muslim organizer, says, "We are trying to bridge the gap between the young working professionals of the DC Green Muslims and local residents who face social and ecological injustices."[9] They learned the importance of coalition building and the difficulty of sustained and consistent involvement partly because of the culture of DC itself—a transient city in which young people and politicos come and go. As interest in the dinners themselves was high and there was a core leadership, a shift occurred when those core leaders moved into other phases of their lives. There was a lull in participation until the efforts of a new crop of DC Green Muslims, led by Sarah Jawaid, came along.

Sarah got the DC Muslims to participate in the *Huffington Post's* No Impact Week. This project gave people the opportunity to examine and reduce their ecological footprint by taking part in a short, intense period of conscious consumption supported by local and online communities. To make No Impact Week relevant to her fellow Muslims, Sarah created an addendum using Qur'anic references to highlight the eco-spiritual ethics in Islam. "As a result of this initiative," she says, "we renewed a sense of community and reinvigorated our efforts to continue creating a space for Muslims to discuss environmental issues guided by spirituality." The following chapters will continue to highlight the stories of Muslims who are significantly involved in the environmental movement and who have dedicated themselves to living a Green Deen. Each person exemplifies one or more of the Green Deen principles and contributes to the movement on an individual, family, and community level.

If you already consider yourself green, can you bring people together to create a sense of fellowship and tighten the bonds of your community around our shared trust (*amana*) with Allah to protect the planet?

Notes

1. *Shamaa-il Timidhi*, trans. Muhammad Bin Abdur Rahman Ebrahim (New Delhi: Adam Publishers), hadith 8.
2. The material in this section is from an interview with Aziz Siddiqi by the author.
3. The information in this section was gathered over the course of several conversations between Sarah Sayeed and the author in the summer and fall of 2009.
4. Comments made at April 22, 2010, Faith Leaders for Environmental Justice Breakfast.

5. Ninth Annual Interfaith Iftar Fast Break, "Diet or Buy It? Faith, Food, and Resource Consumption," co-hosted by the Union Theological Seminary, the Muslim Consultative Network, and the Interfaith Center of New York, in partnership with the Columbia University Muslim Students Association, held at the Union Theological Seminary, September 15, 2009. The panelists were Liore Milgrom-Elcott, associate director of special projects at Hazon; Imam Muhammad Hatim, Admiral Family Circle Islamic Community and Visiting Professor GTS; Naresh Jain, National and International Jain representative; Dr. Hal Taussig, Visiting Professor of New Testament, Union Theological Seminary; and moderator, Ibrahim Abdul-Matin.
6. Ibid.
7. The material in this section is from number of conversations between Sarah Jawaid and the author in October of 2009.
8. All Dulles Area Muslim Society, *Green Environment Guide* (Washington, DC: Adams Center, 2009), AdamsCenter.org.
9. Quotations from Sarah Jawaid and the following information on the DC Green Muslims are from an interview with Sarah Jawaid by the author.

DISCUSSION QUESTIONS

1. Describe the deep symbolism of the color green in Islam.
2. Describe Aziz Siddiqi's pioneering efforts.
3. Give examples of interfaith environmental efforts.
4. Who is Sarah Sayeed?
5. Explain the concept of *khalifah*.
6. What have the DC Green Muslims accomplished?

POLITICS

The Israeli-Palestinian crisis is probably one of the most difficult sociopolitical topics to teach or discuss. On both sides of the crisis, there are vehement supporters who process the situation at opposite ends of the spectrum. The crisis has been going on since before the creation of the nation-state of Israel in 1948, and it cuts profoundly across a host of deeply personal religious and national issues. The tensions are always mounting, but most tragically this crisis is a central narrative that is exploited and distorted by fundamentalist forms of Islam the world over. The following article contains a small synthesis of the controversial practice of suicide attacks within the crisis. Suicide is prohibited in the Qur'an, yet the practice is now easily associated with the crisis as a whole.

ARTICLE 2

ISRAEL AND PALESTINE

By Robert A. Pape and James K. Feldman

Trajectory of Suicide Attacks

Although Israeli occupation of the territories existed since 1967 and Palestinian groups such as Hamas and PIJ had formed in the 1970s and 1980s, we do not see the use of suicide attacks until 1994. Up until that time, Palestinian groups experimented with methods of attack such as guerilla operations and nonsuicide forms of terrorism. The Palestinian suicide campaign is characterized by three periods: 1994 to 1999; 2000 to 2005; and 2006 to the present. The first period is composed of a series of subcampaigns whose success played an important role in subsequent periods. This section will describe in detail when, how, and why suicide attacks have been employed to achieve the Palestinian groups' stated goals in the three campaigns and the subcampaigns within the first phase.

Period No. 1—1994–1999

The first period, from 1994 to 1999, is important because Israel made significant concessions that increased the confidence of the terrorist groups in the coercive effectiveness of suicide attack, and so encouraged the most intense later campaign, the Second Intifada against Israel.

While the adoption of the Oslo Accords in September 1993 led to the decline of ordinary violence in the First Intifada, breakdowns in the Oslo peace process played a key role in the rise of suicide terrorism. The Oslo Accords marked the first face-to-face agreement between Palestinian and Israeli leaders. It entailed a framework for future relations between Israel and the future Palestinian state, and was signed by Yasser Arafat on behalf of the Palestinian Liberation Organization and Israeli Prime Minister Yit-zhak Rabin. The Accords were intended to be a 5-year maximum interim agreement, to allow for resolving major issues such as Palestinian refugees, Jerusalem, security, borders, and Israeli settlements. The most significant features of the Declaration of Principles include

- Creation of the Palestinian Authority (PA) and an elected Council in order to administer Palestinian lands; these lands in the West Bank and Gaza include three types of areas: 1) those allowed full PA control, 2) part PA civilian control and Israeli security control, and 3) Israeli settlements with full Israeli security control and limited PA civilian control. Over the five year-interim period, Israel would grant the PA self-government in phases.
- Withdrawal of the Israeli Defense Forces (IDF) from Gaza and parts of the West Bank
- *Letters of Mutual Recognition*: Israel recognized the PLO as legitimate representation of the Palestinian people while the PLO denounced terrorism and recognized Israel's right to exist.

Subcampaign no. 1—1994 attacks. Shortly after the signing of the Oslo Accords, the IDF was obligated to withdraw from Gaza and Jericho in the West Bank, called the Gaza-Jericho agreement. The IDF did not withdraw by the March 1994 deadline due to failed negotiations between the two sides; Israel wanted to retain its jurisdiction for criminal prosecution and argued for a smaller Palestinian police force. Hamas then conducted two suicide attacks on April 6 and 13 in Israel, which killed 15 Israeli civilians. Members of the Knesset (Israeli legislature) voted to proceed with the Gaza-Jericho agreement on April 18, forgoing its jurisdiction and security requests, and withdrawal began shortly thereafter on May 4, 1994.

The halting of Hamas's suicide subcampaign at this juncture is significant. In the Hebron massacre of February 1994, an Israeli settler killed 29 Palestinians. Thus, Hamas's suicide campaign was originally intended as retaliation for the massacre, as the group announced a series of five planned attacks. However, after the Knesset voted to withdraw from Gaza and Jericho and concede to Palestinian requests, Hamas did not conduct its final three attacks. A statement by Hamas's then political bureau chief, Musa Abu Marzuq, indicates that this type of change in attack plans is not out of line for Hamas: " the military strategy is a permanent strategy that will not change. The modus operandi, tactics, means, and timing are based on their benefit. They will change from time to time."[1] Hamas, having evaluated its actions as accelerating the withdrawal, changed its operations based on the achieved desired outcome. Notably, former Israeli prime minister Yitzhak Rabin also announced a change of plans in response to the attacks: "We have seen by now at least six acts of this type by Hamas and Islamic Jihad. . . . The only response to them and to the enemies of peace on the part of Israel is to accelerate the negotiations."[2] The outcome of the 1994 attacks could only have bolstered Hamas's belief in the coercive value of suicide terrorism; the deadline of the Oslo Accords did not produce change on the ground, but violent action in fact did.

Subcampaign no. 2—1995 attacks. From October 1994 to August 1995, Hamas and Palestinian Islamic Jihad (PIJ) launched a series of nine attacks combined, but this time with the intended goal of Israeli withdrawal from the West Bank. The sequence of events is illustrated by table 6.1.

As illustrated in the table, Israeli withdrawal seems to have been accelerated by violent action on the parts of Hamas and PIJ, supported by the following interpretations of these events. When questioned about the decision to withdraw in the face a potential increase in suicide attacks, Prime Minister Rabin responded, "What is the alternative, to have double the amount of terror? . . . All the bombers were Palestinians who came from areas under our control."[3] In an editorial in the *Jerusalem Post*, the writer warns of the danger of appearing to reward terrorism, "They [the attackers] firmly believe that negotiations will achieve little, and that only bloodshed will cause Israel to withdraw quickly and completely. . . . Unfortunately, the messages they receive from Israeli officials do not tend to discourage them."[4] A statement from Hamas leader Mahmud al-Zahhar reflects this encouragement, "Any fair person knows that the military action was useful for the [Palestinian] Authority during negotiations." Adding to this sentiment, PIJ leader Dr. Ramadan Abdallah Shallah stated, "This is the first time in the history of the Zionist entity

TABLE 6.1 **Action-Reaction**

Action	Reaction
Israel did not withdraw from populated areas of the West Bank by the assigned Oslo Accords deadline of July 1994	Hamas launched three suicide attacks resulting in 30 Israeli deaths
	PIJ launched four attacks resulting in 32 deaths between October 1994 and June 1995
The PA requested a cease fire in March 1995 after Israel agreed to begin withdrawals in July 1995	Hamas and PIJ suspended attacks
Israel withdrawal was delayed due to security road construction setbacks, planned for April 1996 at the earliest	Hamas launched two suicide attacks on July 24 and August 21, 1995, resulting in 11 Israeli deaths
Hamas attacks (aforementioned attacks)	Israel withdrawal was planned and executed on December 1995 without finished construction
	Hamas and PIJ Suicide attacks halted

that martyrdom actions forced the Zionist mind to question the usefulness of establishing a Jewish state on usurped land, on the ruins of a people who will chase them back with resistance and Jihad until the end of time."

Subcampaign no. 3—1996 attacks. The following two Hamas-led sub-campaigns did not result in territorial concessions in the same way as sub-campaigns no. 1 and no. 2, nor were such concessions Hamas's stated goal in either case. With Palestinian aggression on the rise, Israeli counterterrorism efforts were as well, resulting in targeted assassinations of Palestinian group leaders. On October 26, 1995, PIJ leader Fathi Shiqaqi was assassinated by Israeli forces, in addition to the assassination of Hamas leader Yahya Ayyash on January 5, 1996, through a bomb-laden cell phone.[5] Ayyash was responsible for introducing the suicide bombing tactic to Hamas, and was also the head bomb-maker. Following a six-month lull in suicide attacks, Hamas launched four suicide attacks over two weeks in February and March 1996, reportedly to avenge the assassination of one of their most valuable leaders. Shortly after the assassination, the Izz al-Din al-Qassam Brigades of Hamas issued a statement saying, "The martyrdom of the leader, Engineer Yahya Ayyash will only strengthen the Brigades and emphasize its continuation on his path and proceed on the road of Jihad." Moreover, Hamas leader Imad al-Faluji stated that "Israel will pay a high price for this operation."[6] Indeed, the four suicide attacks resulted in 58 Israeli civilian deaths. After the attacks, Palestinian public support for suicide terrorism was at an all-time low of 5%, compared to nearly 80% who opposed it. The PA's crackdown on militant groups combined with unprecedented Palestinian condemnation of the violence upon Israeli civilians led to a year-long lull in suicide attacks.

Subcampaign no. 4—1997 attacks. From March 1997 to September 1997, Hamas carried out three suicide attacks in Israel, resulting in 24 Israeli deaths. After the March and July 1997 attacks, Israel attempted to assassinate Hamas's political bureau chief, Khaled Mash'al, in September 1997 in Amman, Jordan.[7] The failed attempt resulted in the capture of the Israeli agents and a third suicide bombing on September 4. After claiming responsibility for the September bombing, Hamas issued a statement promising to continue attacks until release of Hamas members from Israeli prisons:

> We vow to continue our blows as long as our demands are not met and we will not accept any military truce before our prisoners receive their right to freedom. The attack is our explosive message to the government of the enemy which blockades our people and which destroys our freedom by destroying houses and waging campaigns of oppression and arrests.[8]

As part of a deal between King Hussein of Jordan and Israel, Israel released Hamas leader Sheikh Ahmed Yassin on October 1 in exchange for the captured Israeli agents, in an effort to "respond to King Hussein's appeal and to take positive steps to help the peace process," stated Israeli Prime Minister Benjamin Netanyahu. Sheikh Yassin's release was demanded by Hamas on several occasions, and there is no evidence that the release was interpreted to be a result of Hamas's 1997 suicide attacks. U.S. and Jordanian pressure, combined with Sheikh Yassin's deteriorating health, has been cited as cause for the release.[9] Despite this fact, Hamas did not continue suicide attacks after the release, virtually halting all suicide attacks until the second uprising, or intifada, of 2000.[9] (There are two suicide attacks in our database in 1998, one by Hamas, the other by Islamic Jihad. There were none in 1999.)

Period No. 2—Al-Aqsa Intifada, 2000–2005

The second period, from 2000 to 2005, involved the most severe suicide campaign against Israel, the Second Palestinian Intifada. This campaign was triggered by the evident failure of the Oslo peace process as a result of the Camp David negotiations in August 2000.

The Second Intifada, also known as Al-Aqsa Intifada, differed markedly from the First Intifada of 1987–92. Palestinian resistance moved from unarmed violent rebellion with no use of suicide attacks in the First Intifada, to large-scale suicide bombing and armed rebellion in the Second Intifada. Hamas leader Mash'al explains the tactical transition:

> Like the intifada in 1987, the current intifada has taught us that we should move forward normally from popular confrontation to the rifle to suicide operations. This is the normal development. ... We always have the Lebanese experience before our eyes. It was a great model of which we are proud.[10]

The "Lebanese experience" that Mash'al is referring to is the 1982 Is-raeli invasion into Lebanon. Hezbollah used suicide attacks and violent resistance against U.S. troops after the invasion, and in 1983, U.S. and French troops quit the area. This experience is cited by almost all of the Palestinian groups as proof of the strategic effectiveness of suicide terrorism.

The onset of the Second Intifada came soon after the failure of Camp David negotiations in August 2000 and began a day after Israeli opposition leader Ariel Sharon's controversial visit to the Temple Mount in September 2000, a religious site sacred to both Jews and Muslims. His visit seemed to signal the overall collapse of the Oslo peace process. In the Camp David negotiations U.S. President Bill Clinton invited Palestinian leader and Palestinian Authority Chairman Yasser Arafat and Israeli Prime Minister Ehud Barak to end the overextended five-year interim period of the Oslo Accords and resolve major issues toward a final settlement of the Israeli-Palestinian conflict. Ultimately, no agreement was reached and leaders instead signed a trilateral statement that defined agreed principles to guide future negotiations. The principles simply stated that major issues should be resolved as soon as possible and with good-faith intentions. Both sides blamed the other for the failure of the Summit to reach closure. Loss of confidence in the peace process ultimately led to the Second Intifada, or uprising, of 2000.

All four of the groups—Hamas, PIJ, Al-Aqsa, and PFLP—carried out suicide terrorism during the intifada, with a combined total of 106 suicide attacks. The sudden rise and persistence of suicide attacks on the part of the Palestinian groups can be explained by three key developments: 1) the increasing and punctuated disillusionment with salient negotiation processes, 2) the public shift in acceptance and support of suicide terrorism, and 3) the decreasing legitimacy of the PA and Fatah party under Yasser Arafat. In light of these events, the Palestinian groups increasingly employed suicide attacks, viewing this method as a last resort. Munir al-Makdah, suicide bombing trainer, explains this transition: "Jihad and the resistance begin with the word, then with the sword, then with the stone, then with the gun, then with planting bombs, and then transforming bodies into human bombs." From this quote and Mash'al's quote above, we can see that the Palestinian groups viewed this transition as a strategic choice of attack.

Like Prime Minister Yitzhak Rabin in the mid-1990s, Prime Minister Ariel Sharon's hardline government, which took office in March 2001, ultimately responded to the suicide attacks with territorial concessions. At first Sharon relied on heavy ground forces. In 2002, Israel launched "Operation Defensive Shield," reoccupying several Palestinian cities and towns, constructing a security fence, and pursuing targeted leader assassinations in order to destroy terrorist infrastructure. After these efforts, the number of attacks decreased from 42 in 2002 to 19 in 2003, and fatalities decreased from 250 in 2002 to 126 in 2003. However, the threat of suicide terrorism remained significant.

What brought an end to the Al-Aqsa Intifada was Israel's unilateral disengagement plan, which was announced by Ariel Sharon on December 18, 2003. He stated: "It is not in our interest to govern you. We will not remain in all the places where we are today." He also stated that he would give them back the Gaza Strip by evacuating all Jewish settlements there, withdraw Israeli military forces from a large part of the West Bank, and complete a security barrier between Israel

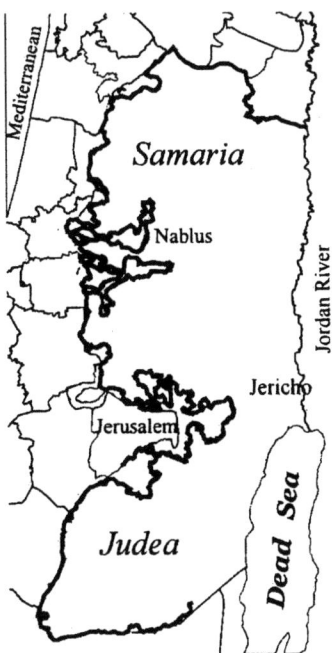

Figure 6.1. Israeli West Bank barrier

and the abandoned territory on the West Bank, signaling a commitment to keep Israeli forces out of the West Bank as much as a commitment to keep suicide bombers out of Israel.[11] The triple fence security barrier was mostly completed by late 2004 after which Israel's withdrawal from Gaza and part of the West Bank began. After the withdrawal process ended in September 2005, suicide terrorism attacks almost completely stopped. Suicide attacks noticeably began to decline just a few months following Sharon's announcement.

Sharon's unilateral withdrawal decision is widely viewed as a response to suicide attacks, a view shared by many Israelis as well as Palestinians. For example, on July 18, 2005, *Haaretz's* Danny Rubinstein wrote, "Sharon, who never once mentioned or alluded to the need to withdraw from Gaza before, needed suicide bombers, rockets, and mortars to persuade him."[12]

Period No. 3—2006–Present

Despite fears of a resurgence of suicide terrorism following Hamas' electoral victory in January 2006, a third period of very low levels of suicide terrorism continues to this day. The third period, from 2006 to the present, is characterized by a significant drop in suicide attacks and coincides with Israeli disengagement from Gaza and large parts of the West Bank. The vast decline of the Palestinian suicide attacks in response to Israeli withdrawal reinforces the overall patterns that this method of attack is driven mainly by strategic concerns.

Notes

1. Fatah charter, http://www.mideastweb.org/fateh.htm.
2. "Palestinian Fatah Armed Wing Says Negotiations with Israel 'Useless,'" *BBC Monitoring Middle East, August* 17, 2003.
3. Shaul Mishal and Avraham Sela, *Palestinian Hamas: Vision, Violence, and Coexistence* (New York: Columbia University Press, 2000).
4. David Makovsky and Alon Pinkas. "Rabin: Killing Civilians Won't Kill the Negotiations," *Jerusalem Post*, April 13, 1994.
5. Yitzhaq Rabin, "Interview," *BBC Summary of World Broadcasts*, September 8, 1995.
6. "Suicide Bombers Return," *Jerusalem Post*, November 3, 1995.
7. "Gaza: 100,000 Palestinians Protest Assassination," *The Militant*, January 22, 1996.
8. "Israel to Pay 'High Price' for Ayyash Killing," *Radio Monte Carlo*, Paris, France, January 6, 1996.
9. Anne Marie Oliver and Paul Steinberg, *The Road to Martyrs' Square: A Journey to the World of the Suicide Martyr* (New York: Oxford University Press, 2005), p. 45.
10. "HAMAS Claims Responsibility for Jerusalem Blast, Threatens More," *Agence France Presse*, September 4, 1997.
11. Serge Schmemann, "Israel Frees Ailing Hamas Founder to Jordan at Hussein's Request," *New York Times*, October 1, 1997, p. A5.
12. It should be noted that violent attacks in the form of car bombs and planted bombs did occur between 1997 and 2000, but suicide attacks did not.
13. "Hamas Statement," *BBC Summary of World Broadcasts*, November 3, 2001.
14. Azzam Tamimi, *HAMAS a History from Within* (Northampton: Olive Branch Press, 2007), p. 205.
15. Danny Rubinstein, "Palestinian Pride, Israeli Capitulation," *Haaretz*, March 21, 2005.

DISCUSSION QUESTIONS

1. When did suicide attacks begin? List the dates of the three distinct periods.
2. Describe the mission of the Oslo Accords, and summarize the Declaration of Principles.
3. What led to the first set of suicide attacks?
4. Describe the period of the Al-Aqsa Intifada.
5. What happened to the rate of suicide attacks over time?
6. Briefly describe the period we are in now.

WOMEN

Much has been said about Muslim women, who unfortunately are sometimes presented as being oppressed and second-class citizens; it is regrettable that their voices have often been left out of Western scholarly conversations. However, within the last few decades Muslim women, even Muslim feminists, have been able to attract an international audience. They have come out resoundingly as strong, intelligent, hardworking leaders, and they see Islam as a big reason why that is true. Many Muslim women feel that Islam dignifies them and gives them the platform to show their strength. Yet many other Muslim women are still struggling to simply obtain what they understand to be basic human rights. The following article examines some of these rights.

ARTICLE 3

RESTRICTIONS OF THE RIGHTS OF WOMEN

By Ann Elizabeth Mayer

Women's Rights in Pakistan

The enormously complicated situation in Pakistan defies easy characterization. Although Pakistan is one of the rare Muslim countries to have elected a woman as prime minister, powerful Islamist factions agitate constantly for restrictions on women's rights, and many support the Taliban model of subjugating women. Pakistan's assertive feminists, including the distinguished human rights lawyers Asma Jahangir and Hina Gilani, mobilize and litigate, seeking to thwart proposals for more Islamization and to roll back discriminatory rules and practices. At the same time, the authorities have a record of condoning rape, tolerating egregious incidents of domestic violence, abusing women held in detention, and failing to punish perpetrators of honor killings. It is important, however, to note that many of the most notorious incidents of violence against women in Pakistan have no connection to Islam, stemming instead from retrograde local customs and mentalities imbued with primitive sexism.

Not surprisingly, there were contentious debates about whether Pakistan should ratify the Women's Convention. Ultimately, a compromise was made; the 1995 ratification was said to be "subject to the provisions of the Constitution of Pakistan." Because, as previously noted, the constitution called for Muslims to be enabled to live in "accordance with the teachings and requirements of Islam," what appeared to be a "constitutional" reservation was potentially the equivalent of the Islamic reservations that had been entered by other Muslim countries.

In a striking departure from the typical pattern of using Islamic law to curb rights and freedoms since the inauguration of Pakistan's Islamization program, in 2006 Pakistan's Federal Shariat Court referred both to the Islamic sources and international law in an opinion declaring section 10 of the Pakistani Citizenship Act, 1951, to be discriminatory against women.[1] The citizenship rules allowed a foreign woman marrying a Pakistani man to obtain Pakistani citizenship but not a foreign man marrying a Pakistani woman. When the rules were challenged, the government took the position that there was a threat of men from other countries misusing the opportunity to gain Pakistani citizenship through marriage, especially Afghan refugees and illegal Bengali and other South Asian immigrants. It warned that if women could obtain Pakistani citizenship for their foreign husbands, such undesirables could marry Pakistani women and after getting citizenship divorce them.[2] The court was not impressed and said of the section that it was:

> discriminatory, negates gender equality and is in violation of Articles 2-A (Objective Resolution) and 25 (equality of citizens) of the Constitution, also against international commitments of Pakistan and, most importantly, is repugnant to the Holy Quran and Sunnah.[3]

Among other comments, the court asserted: "The last sermon of Holy Prophet is the first Charter of Human Rights wherein all human beings are equal."[4] That is, in this unusual and progressive court opinion, the Shariat Court offered a different way of thinking about Islam and human rights, treating the Islamic sources as supporting international law and upholding women's equality, thereby reminding Pakistanis that Islam did not have to be treated as mandating an inferior status for women. Here Islam was used as a tool to discredit a discriminatory secular law. The court requested the President to amend the law within six months. Regrettably, years later, the change in the law had still not gone through.[5]

The New Afghan and Iraqi Constitutions

In the drafting of the most recent constitutions of Afghanistan and Iraq, pressures for Islamization have been countered by demands for protecting women's rights. In the postinvasion era, the influence of US officials in both countries added an extra factor to the resistance to Islamization. Nonetheless, Islamic elements were included in both documents that could be exploited to curb women's rights.

In the turbulent, war-torn years just before the Taliban takeover, various factions competed for supremacy in Afghanistan. During this period, women suffered from abduction, displacement, abuse, torture, rape, and slaughter.[6] All this was but a prelude to the even harsher crackdowns and more extensive abuses perpetrated under the Taliban, who succeeded in extending their domination over most of the country in 1996. They unleashed fierce enforcers of their retrograde version of Islamic morality to terrorize the population into submission. Ironically, their Islamic dress rules were less clearly discriminatory than many others because men, like women, were required to don approved Islamic dress. Women had to be completely swathed in the enveloping burqa when in public, and their mobility was sharply curbed because they were not allowed go out without a male relative to escort them. The Taliban's policies amounted to a regime of gender apartheid, barring women from all education and virtually all employment outside the home and even blocking women's access to health care.[7]

Seemingly chastened by the international chorus of condemnation and the refusal of most states to recognize their regime, Taliban officials denied charges that they were trampling on women's human rights.[8] Their denial proved that the prestige of human rights had grown to the point where even regimes pursuing the most retrograde Islamization programs felt obliged to deflect charges of violations.

Afghan women struggled against the Taliban's program and mobilized to support the opposition.[9] Many celebrated the Taliban's 2001 downfall and the adoption by the new government of policies more supportive of women's rights. Women pressed hard to have protections for their rights written into the 2003 Afghan constitution, but they came out of the difficult drafting process with only limited victories; the constitution has both worrisome and positive features. It advises in Article 3 that no law should contravene the beliefs and principles of the sacred religion of Islam, which could be used to bar laws advancing women's rights. At the same time, Article 22 provides that citizens, whether men or women, have equal rights and duties before the law, which could ban discrimination against women if "the law" being referred to were not itself discriminatory. Moreover, the post-Taliban government ratified the Women's Convention in 2003 without imposing any Islamic reservations, which seemed a positive step.

Meanwhile, Afghan women in the aftermath of the US invasion were forced to cope with turmoil and violence, as a weak central government struggled unsuccessfully to assert control beyond the capital and to uproot vestiges of the Taliban in the war-ravaged, economically prostrate country.[10] By 2011, Afghanistan ranked in one survey as the most dangerous country for women, more dangerous even than Congo or Somalia.[11] In these circumstances, the promise of eliminating discrimination against women and improving their lives seemed little more than an idealistic illusion, and boasts about the US mission of "saving" Afghan women could be viewed as the hypocritical rhetoric of "the new colonial feminism."[12]

The situation in Iraq was similarly chaotic. Although the ravages of Saddam Hussein's despotism affected all Iraqis, women fared relatively well under his dictatorship when compared with their sisters in Iran and Afghanistan. A reformed version of Islamic law prevailed in the area of personal status, and many opportunities were open to women. The US invasion displaced Saddam's Ba'athist ruling clique and enabled long-suppressed Shi'i forces to assume power, creating a situation where Islamist extremists could exert pressures on women to veil themselves and retire to the domestic sphere. Various Shi'i and Sunni factions, as well as more secular groups and Kurds, struggled bitterly but inconclusively to realize conflicting agendas. The huge increase in crime and terrorist violence in the chaos that ensued created such perils for women that many felt compelled to stay locked inside their homes or to seek refuge in other countries.

The United States publicly pressed the drafters of the Iraqi constitution to include an express guarantee of equality for women.[13] (In part because of US pressures, the 2004 Iraqi Transitional Law, in Article 12, had already barred discrimination "on the basis of gender.") Nevertheless, ambiguous provisions in the 2005 Iraqi constitution left women's rights in doubt. Article 14 provides that all Iraqis are equal before the law without discrimination because of gender, race, ethnicity, origin, color, religion, creed, belief or opinion, or economic and social status, and Article 20 gives male and female citizens political rights, including the right to vote and run for office. A different impression is created by provisions in Article 2 that state that no law may contradict the undisputed rules of Islam and that Islamic law is to be a main source of legislation. These provisions open the door to the application of Islamic criteria in conflict with women's rights. Furthermore, Article 39 contains a vague provision allowing Iraqis to be governed by religious law in matters of personal status. This would allow men to ask the courts to impose premodern Islamic jurisprudence with all its patriarchal features, an outcome that troubled Iraqi feminists. Taken together, these provisions suggest that the constitution left room for deploying Islamic rules at the expense of Iraqi women's rights, especially in personal status matters.

The Influence of Sex Stereotyping

CEDAW recognizes that sex stereotyping constitutes an obstacle to realizing full equality for women and calls on governments to attack the attitudes and practices that stereotype women as inferior beings whose nature disqualifies them from enjoying freedoms on a par with men. Article 5.a binds the parties "to modify the social and cultural patterns of conduct of men and women, with a view to achieving the elimination of prejudices and customary and all other practices which are based on the idea of the inferiority or the superiority of either of the sexes or on stereotyped roles for men and women."

One sees no concern for transcending or eliminating sex stereotyping in the Islamic human rights schemes reviewed here; on the contrary, such stereotyping is a central feature of the schemes, if sometimes only an implied one. The belief that men and women have inherently different natures and thus have distinct rights and obligations is reflected in the formal reservations to CEDAW expressed by several Muslim countries, which combine references to Islam and assumptions that women and men must play different, complementary roles.[14] According to proponents of the complementarity thesis, the goal should be equity that recognizes differences between men and women, not equality in rights that disregards such differences.

Sultanhussein Tabandeh freely expresses his stereotypical views of women in the course of explaining why the human rights accorded to women in the UDHR are incompatible with Islam. Women, according to Tabandeh, are touchy and hasty, volatile and imprudent. They are generally more gullible and credulous than men. Their sexual desire makes them easy prey for the blandishments of salacious individuals.[15] In Tabandeh's opinion, women were designed for "cooking, laundering, shopping, and washing up," as well as for taking care of children. Men, in contrast, were created for field work, warfare, and earning a living.[16] Women are deficient in the intelligence needed for "tackling big and important matters"; they are prone to making mistakes and lack long-term perspective. For this reason, he said, they must be excluded from politics.[17] Women cannot fight in war because they are "timorous-hearted," limited by physical weakness, and may become frightened and run away.[18]

Abu'l A'la Mawdudi takes a similar line, although not in his publication on human rights. In his book on purdah, he argues that nature has designed men and women for different roles, treating menstruation, pregnancy, and nursing as incapacitating disabilities.[19] Women, he asserts, are created to bear and rear children. They are tender, unusually sensitive, soft, submissive, impressionable, and timid. They lack firmness, authority, "cold-temperedness," strong willpower, and the ability to render unbiased, objective judgment.[20] Men have coarseness, vehemence, and aggressiveness, which make them suited to assume roles as generals, statesmen, and administrators. The education of men should, therefore, aim at training them so that they can support and protect the family, whereas a woman should be educated to bring up children, look after domestic affairs, and make home life "sweet, pleasant, and peaceful."[21]

Similar observations were made by Ayatollah Javad Bahonar, a cleric who briefly served as Iran's prime minister before being assassinated. Bahonar was one of Khomeini's closest aides, and his thinking may be taken as representative of many of the leading clerics in Khomeini's regime. In an article on Islam and women's rights in an English-language journal distributed in the West by the Iranian regime, Bahonar said that men are bigger and stronger and have larger brains, with a larger proportion of the brain "dealing with thought and deliberation."[22] A relatively larger proportion of the smaller female brain is "related to emotions," and women have more in the way of the affection and deep tender sentiments that suit them for child care and nursing. Women's sentiments and emotions make them ill equipped to cope with earning a livelihood, which calls for farsightedness, perseverance, strength, tolerance, coolness, planning ability, hard-heartedness, connivance, and the like—characteristics that women lack. Men are designed by nature to deal with "the tumult of life," to fight on the battlefield, and to manage the affairs of government and society.[23] A note at the end of the article offers some statistics on female physical inferiority, including the comment that "a man's brain weighs 100 grams more than a woman's." Bahonar summed up his evidence by saying that the "differences in physical structure are reflected in the mental capacities of the two sexes."[24]

Notwithstanding this recital of women's natural deficiencies and infirmities, Bahonar sought to maintain the fiction that Iran's official Islam supported women's rights, asserting that "Islam considers men and women equal as far as the basic human rights are concerned."[25] Although Bahonar cited many ways that men and women share equal religious and moral duties—both are required to pray, to be faithful and obedient believers, to command the good and prohibit the evil, to keep their looks cast down, and to accept punishment for crimes—Bahonar cited only two rights that the two sexes share on an equal basis: the right to own and use property and the right to inherit.[26] On the latter, he neglected to mention that although women have a right to inherit, they take only one-half the share of men inheriting in the same capacity.

A 2003 study of more recent sex stereotyping in religious discourse in Iran notes that it assumes that men and women have opposite characteristics. Among other things, men are described as logical and disciplined, women as confused and lacking in discipline; men as courageous, women as fragile; men as authoritarian and independent, women as obedient and in need of protection; and men as not tolerating physical suffering, women as managing to endure physical abuse from their husbands.[27]

Because the Islamic legal tradition developed in traditional, patriarchal milieus and because the authoritative works on the legal status of women in Islam have been exclusively written by men, it is not surprising that men's stereotypes of women were incorporated and that they have been retained as part of Islamic rights schemes. In contrast, one finds little in the original Islamic sources that supports these stereotypes, and there is, moreover, nothing distinctively Islamic about the self-interested and biased appraisals of women's inherent traits that they offer. In fact, the sex stereotyping that one sees in the Islamic tradition resembles stereotypes found in other cultural and religious contexts. For example, the Saudi prohibition against women driving, which is associated with conservative Islamic precepts, has a historical counterpart in the United States of the early twentieth century, when sex stereotyping was used to justify prohibiting women from driving cars.[28]

The Catholic Church is only one of many denominations where sex stereotyping has been used to support the view of church clergy—all men—that women must be kept subjugated because of their "natural" inferiority to men. Among the characteristics that the church fathers attributed to women were fickleness, shallowness, garrulousness, weakness, slowness of understanding, and instability of mind.[29] Saint Augustine asserted that compared to men, women were small of intellect.[30] In a 1966 book designed to persuade Catholic women that because of their gender, their "entire psychology is founded upon the primordial tendency to love," two priests asserted that a woman's brain "is generally lighter and simpler than man's."[31]

Pope Pius XII, speaking to an audience of women in 1945, described "the sensibility and delicacy of feeling peculiar to woman, which might tempt her to be swayed by emotions and thus blur the clearness and breadth of her view and be detrimental to the calm consideration of future consequences."111 He maintained that as a result of their characteristics, women were suited for tasks in life that called for "tact, delicate feelings, and maternal instinct, rather than administrative rigidity."[32]

As a major feminist critic of Catholic doctrine has pointed out, the sex stereotypes upheld by men in the church hierarchy became closely intertwined with the Catholic teachings calling for the subordination of women: "The very emancipation which would prove that women were not 'naturally' defective was denied them in the name of that defectiveness which was claimed to be natural and divinely ordained."[33]

One can see the same presuppositions about inherent female characteristics in both the Islamic and Christian traditions. The question that remains is whether the sex stereotypes associated with religious doctrine are actually supported by the original sources or whether they are simply being read into religious teachings by male interpreters with patriarchal biases.

Summary

The failure to accord women equality in rights turns out to be one of the ways that Islamic human rights schemes deviate most dramatically from international human rights law. The intention to accommodate discriminatory rules is cloaked by vague Islamic conditions placed on women's rights that may look harmless to a casual observer but that, depending on how they are interpreted, have the potential to nullify women's rights. In the wake of the upheavals of the Arab Spring, a number of countries will be wrestling with formulating rights provisions in their new constitutions. With the dramatic rise of Islamist political forces, it is entirely possible that women who fought and sacrificed alongside men in the dangerous struggles to defeat dictatorships will have to mobilize to combat projects aimed at restricting their rights under the rubric of upholding Islamic law. Knowing how Islamic conditions on women's rights in documents such as the Iranian constitution have correlated with official efforts to demote women to second-class status, they will be alert to the implications of constitutional provisions that accord priority to upholding Islamic law. It is sobering to observe that, after many decades of campaigning to secure their equality, women in the Middle East may have reason to worry lest they see a repetition of the loss of freedoms that Iranian women endured in the wake of a revolution that many had hoped would usher in a new dawn of freedom.

Notes

1. The reforms made by the Qur'an are presented schematically in a volume produced under the auspices of the Giant Forum and Global Issues Awareness for National Trust in collaboration with the Women's Development Fund, Canadian International Development Agency (CIDA), Islamabad, Pakistan. See *International Conference on Islamic Laws and Women in the Modern World: Islamabad, December 22–23, 1996* (Islamabad: Giant Forum, 1996), 20–21. Hereafter, cited as *International Conference on Islamic Laws*.
2. Fazlur Rahman, "The Status of Women in the Qur'an," in *Women and Revolution in Iran*, ed. Guity Nashat (Boulder: Westview Press, 1983), 38.
3. Jane Smith, "Women, Religion, and Social Change in Early Islam," in *Women, Religion, and Social Change*, eds. Yvonne Haddad and Ellison Findley (Albany: State University of New York Press, 1985), 19–35.
4. See the examination of disparities between the original sources and later interpretations in Barbara Stowasser, "The Status of Women in Early Islam," in *Muslim Women*, ed. Freda Hussain (New York: St. Martin's Press, 1984), 11–43; Asma Barlas, *"Believing Women" in Islam: Unreading Patriarchal Interpretations of the Qur'an* (Austin: University of Texas Press, 2002); and Fatima Mernissi, *The Veil and the Male Elite: A Feminist Interpretation of Women's Rights in Islam* (Reading, MA: Addison-Wesley, 1991).

5. Rahman, "The Status of Women," 37.

6. Introductions to aspects of women's status in the *shari'a* can be found in Joseph Schacht, *Introduction to Islamic Law* (Oxford: Clarendon Press, 1964), 126–127; Yves Linant de Bellefonds, *Traité de droit musulman comparé*, vol. 2, *Le Mariage: La Dissolution du mariage* (Paris: Mouton, 1965); Noel Coulson, *Succession in the Muslim Family* (Cambridge: Cambridge University Press, 1971); Ghassan Ascha, *Du statut inférieur de la femme en Islam* (Paris: L'Harmattan, 1987); and *International Conference on Islamic Laws*.

7. A summary of these changes can be found in J. N. D. Anderson, *Law Reform in the Muslim World* (London: Athlone, 1976). See also Tahir Mahmood, *Personal Law in Islamic Countries* (New Delhi: Academy of Law and Religions, 1987); and *International Conference on Islamic Laws*, 185–453.

8. A perfect embodiment of this response can be found in Abu'l A'la Mawdudi, *Purdah and the Status of Women in Islam* (Lahore: Islamic Publications, 1979). Many aspects of this literature are reviewed by Ascha, *Du statut inférieur*.

9. For a learned discussion of contraception and abortion in Islamic jurisprudence, see Basim Musallam, *Sex and Society in Islam: Birth Control Before the Nineteenth Century* (Cambridge: Cambridge University Press, 1983).

10. See Abdullahi El-Naiem [An-Na'im], "A Modern Approach to Human Rights in Islam: Foundations and Implications for Africa," in *Human Rights and Development in Africa*, eds. Claude Welch Jr. and Ronald Meltzer (Albany: State University of New York Press, 1984), 82; and Mernissi, *The Veil and the Male Elite*.

11. Works that show the growth and influence of feminism include Mernissi, *The Veil and the Male Elite*; Riffat Hassan, "Feminist Theology: The Challenges for Muslim Women," *Critique: Journal for Critical Studies of the Middle East* (Fall 1996): 53–66; Mahnaz Afkhami, ed., *Faith and Freedom: Women's Human Rights in the Muslim World*; Mahnaz Afkhami and Erika Friedl, eds., *Muslim Women and the Politics of Participation: Implementing the Beijing Platform* (Syracuse, NY: Syracuse University Press, 1997); Ziba Mir-Hosseini, *Islam and Gender: The Religious Debate in Contemporary Iran* (Princeton: Princeton University Press, 1999), and *Feminism and the Islamic Republic: Dialogues with the Ulema* (Princeton: Princeton University Press, 1999); Shaheen Sardar Ali, *Gender and Human Rights in Islam and International Law* (The Hague: Kluwer Law International, 2000); Val Moghadam, "Islamic Feminism and Its Discontents: Toward a Resolution of the Debates," *SIGNS: Journal of Women in Culture and Society* 27 (2002): 1136–1171; Khaled Abou El

Fadl, *Speaking in God's Name: Islamic Law, Authority and Women* (Oxford: Oneworld Publications, 2003).

12. This is a consistent theme of Mawdudi's writings. For an example of his arguments, see Mawdudi, *Purdah and the Status of Women*, 21–24.

13. Ibid., 73–74.

14. Ibid., 24.

15. Thus, in most Muslim countries, the choice has been to compromise, keeping some *shari'a* rules but including many reforms improving the rights of women. See *International Conference on Islamic Laws*, 185–453; and Lynn Welchman, *Women and Muslim Family Laws in Arab States: A Comparative Overview of Textual Development and Advocacy* (Amsterdam: Amsterdam University Press, 2007).

16. See Jane Connors, "The Women's Convention in the Muslim World," in *Human Rights as General Norms and a State's Right to Opt Out: Reservations and Objections to Human Rights Conventions*, ed. J. P. Gardner (London: British Institute of International and Comparative Law, 1997), 85–103.

17. See, for example, Rana Husseini, "Women Activists Welcome Endorsement of Government Decision to Lift Reservation on CEDAW Article," *Jordan Times*, May 20, 2009, available at http://www.jordantimes.com/?news=16860; "Morocco With-draws Reservations to CEDAW," *Magharebia/ADFM*, December 18, 2008, available at http://www.wluml.org/node/4941.

18. Hossam Bahgat and Wesal Alfi, "Sexuality Politics in Egypt," in *SexPolitics. Reports from the Front Lines*, eds. Richard Parker, Rosalind Petchesky, and Robert Sember, 58–62, available at http://www.sxpolitics.org/frontlines/book/pdf/sexpolitics.pdf.

19. Ann Elizabeth Mayer, "Rhetorical Strategies and Official Policies on Women's Rights: The Merits and Drawbacks of the New World Hypocrisy," in *Faith and Freedom*, ed. Afkhami, 105–119, and "Religious Reservations to CEDAW: What Do They Really Mean?" in *Religious Fundamentalism and the Human Rights of Women*, ed. Courtney Howland (New York: St. Martin's Press, 1999), 105–116.

20. Sultanhussein Tabandeh, *A Muslim Commentary on the Universal Declaration of Human Rights*, trans. F. J. Goulding (Guildford, England: F. J. Goulding, 1970), 1.

21. See, for example, his comments on Article 16 of the UDHR. Ibid., 41–45.

22. Ibid., 40.

23. Ibid., 58.

24. Ibid., 51.

25. Abu'l A'la Mawdudi, *The Islamic Law and Constitution* (Lahore: Islamic Publications, 1980), 262–263; and *Purdah and the Status of Women*, passim, and on divorce, 151.
26. Mawdudi, *Purdah and the Status of Women*, 12.
27. Ibid., 12–15, 26–71.
28. Ibid., 15.
29. Ibid., 73.
30. Abu'l A'la Mawdudi, *Human Rights in Islam* (Leicester, England: Islamic Foundation, 1980), 18.
31. Mawdudi, *Human Rights*, 18.
32. Indeed, restrictions on women's testimony in the *shari'a* rules of evidence can make it especially difficult for a woman to prove such an offense.
33. Naturally, the sordid details of these incidents were publicized in India, Pakistan's enemy. See, for example, Amita Malik, *The Year of the Vulture* (New Delhi: Orient Longman, 1972). Mawdudi and his followers had backed the efforts of the Pakistani government to crush the movement to establish an independent Bangladesh, and he preferred to pretend that no rapes had resulted.
34. According to the preface, Mawdudi's human rights pamphlet is translated from a speech delivered on November 16, 1975, at the Civil Rights and Liberties Forum in the Flatties Hotel in Lahore. Mawdudi, *Human Rights*, 7. Mawdudi and his audience in Lahore, capital of the Punjab, which has traditionally supplied most of Pakistan's military manpower, must have been aware of how the Bengali mass rapes had tarnished Pakistan's image.
35. See Shahla Haeri, "The Politics of Dishonor: Rape and Power in Pakistan," in *Faith and Freedom*, ed. Afkhami, 161–174; and Rubya Mehdi, "The Offence of Rape in the Islamic Law of Pakistan," *Women Living Under Muslim Laws*: Dossier 18 (July 1997): 98–108. The sexism of the Pakistani laws on rape in the wake of Islamization is dissected in Asifa Quraishi, "Her Honor: An Islamic Critique of the Rape Laws of Pakistan from a Woman-Sensitive Perspective," *Michigan Journal of International Law* 18 (1997): 287–320.
36. According to the choice-of-law rules in Islam and in the law of Muslim countries, Islamic criteria are used to judge the validity of a mixed marriage. See Klaus Wahler, *Interreligiöses Kollisionsrecht im Bereich privatrechtlicher Rechtsbeziehungen* (Cologne: Carl Heymanns Verlag, 1978), 157–158.
37. Examples of such rules are discussed in Mahmood, *Personal Law*, 275–276.
38. These references appear on p. 19 of the English version.

39. For example, see Mawdudi's invocation of this verse; Mawdudi, *Purdah and the Status of Women*, 149.

40. Yves Linant de Bellefonds, *Traité de droit musulman comparé*, vol. 3, *Filiation: Incapacités, Liberalités entre vifs* (Paris: Mouton, 1965), 81–142.

41. In theory, the guardian could also marry off a male ward without his consent, but because a Muslim man can easily terminate an unwanted marriage, this has little practical significance.

42. Anderson, *Law Reform in the Muslim World*, 102–105.

43. Robert F. Worth, "Tiny Voices Defy Child Marriage in Yemen," *The New York Times*, June 29, 2008, available at http://www.nytimes.com/2008/06/29/world/middle east/29marriage.html?pagewanted=all.

44. See, for example, Mawdudi, *Purdah and the Status of Women*, 144–155.

45. Noel Coulson, *Succession in the Muslim Family* (Cambridge: Cambridge University Press, 1971), 214.

46. Linant de Bellefonds, *Traité de droit musulman comparé*, vol. 2, 451–470.

47. The translation is from "Constitution of the Islamic Republic of Iran of 24 October 1979 as Amended to 28 July 1989," in *Constitutions of the Countries of the World*, eds. Albert Blaustein and Gisbert Flanz (Dobbs Ferry, NY: Oceana, 1992).

48. Eliz Sanasarian, *The Women's Rights Movement in Iran: Mutiny, Appeasement, and Repression from 1900 to Khomeini* (New York: Praeger, 1982), 94–97. The situation of women in the aftermath of the suspension of the Family Protection Act is examined in Ziba Mir-Hosseini, *Marriage on Trial: A Study of Islamic Family Law. Iran and Morocco Compared* (New York: I. B. Tauris, 1993).

49. Shahla Haeri, *Law of Desire: Temporary Marriage in Iran* (London: I. B. Tauris, 1989).

50. *International Conference on Islamic Laws*, 316.

51. Parvin Paidar, *Women and the Political Process in Twentieth-Century Iran* (Cambridge: Cambridge University Press, 1995), 303–335.

52. Ibid., 286–289.

53. See Ziba Mir-Hosseini, "The Politics and Hermeneutics of Hijab in Iran: From Confinement to Choice," *Muslim World Journal of Human Rights* 4 (2007), 1–19, available at http://www.bepress.com/mwjhr/vol4/iss1/art2/.

54. See UN High Commissioner for Human Rights, "Situation of Human Rights in the Islamic Republic of Iran: Iran (Islamic Republic of) 15/10/97," A/52/472, available at http://www.iran.org/humanrights/UN971015.htm.

55. "Iranian Team Set for Atlanta Despite Visa Bother," *Deutsche Presse-Agentur*, July 9, 1996, available in LEXIS/Nexis Library, ALLWLD File.

56. Nikki Keddie, "Women in Iran Since 1979," Special Issue, Iran Since the Revolution, *Social Research* 67 (Summer 2000): 417–419.

57. See Ann Elizabeth Mayer, "Islamic Rights or Human Rights: An Iranian Dilemma," *Iranian Studies* 29 (Summer-Fall 1996): 284–288.

58. "Iranian Cleric Blasts Taleban for Defaming Islam," *Reuters North American Wire*, October 4, 1996, available in LEXIS/Nexis Library, ALLWLD File.

59. "Iranian Leader Warns Women Against Copying Western Feminist Trends," *Agence France Presse*, October 22, 1997, available in LEXIS/Nexis Library, ALLWLD File.

60. Mahsa Sherkarloo, "Iranian Women Take on the Constitution," MERIP Online, July 21, 2005, available at http://merip.org/mero/mero072105.html.

61. See Mehrangiz Kar, Ludovic Trarieux Prize Winner 2002, available at http://www.ludovic-trarieux.org/uk-pages3.1.plt.htm.

62. See Mehrangiz Kar, "Iranian Law and Women's Rights," *Muslim World Journal of Human Rights* (2007): 1–13, available at http://www.bepress.com/mwjhr/vol4/iss1/art9/.

63. See Noushin Ahmadi Khorasani, *Iranian Women's One Million Signatures Campaign for Equality: The Inside Story* (Women's Learning Partnership Translation Series, 2009), available at http://www.learningpartnership.org/iran-oms; Iran's One Million Signatures Campaign, available at http://learningpartnership.org/iran-oms.

64. See "Women Ejected by Force from Iran Stadium," *Iran Focus*, March 6, 2006, available at http://www.iranfocus.com/modules/news/article.php?storyid=6091.

65. Saeed Kamali Deghan, "Iran Jails Director Jafar Panahi and Stops Him Making Films for 20 Years," *The Guardian*, December 20, 2010, available at http://www.guardian.co.uk/world/2010/dec/20/iran-jails-jafar-panahi-films.

66. "Mahmoud Ahmadinejad Blasts Fifa 'Dictators' as Iranian Ban Anger Rises," *The Guardian*, June 7, 2011, available at http://www.guardian.co.uk/football/2011/jun/07/iran-anger-ahmadinejad-fifa-ban/.

67. See Sarah Menkedick, "Iran Wins Membership to the U.N Commission on the Status of Women, Women's Rights," May 08, 2010, change.org, available at http://news.change.org/stories/iran-wins-membership-to-the-un-commission-on-the-status-of-women; Elizabeth Weingarten, "At Last Minute, East Timor Beats Out Iran for Chair of New UN Body," *The Atlantic*, November 10, 2010, available at http://www.theatlantic.com/international/

archive/2010/11/at-last-minute-east-timor-beats-out-iran-for-chair-of-new-un-body/66397/.

68. *The Meaning of the Glorious Koran*, trans. Marmaduke Pickthall (Albany: State University of New York Press, 1976).

69. See, for example, Tabandeh, *A Muslim Commentary*, 51–52; and Mawdudi, *Purdah and the Status of Women*, 185–201.

70. A critique of how the concept of *'ird* has been used to deny Arab women their humanity and basic rights can be found in Nawal El-Saadawi, *The Hidden Face of Eve: Women in the Arab World* (Boston: Beacon Press, 1981), 7–90.

71. The other two countries that recognized the Taliban government were Pakistan and the United Arab Emirates.

72. Declarations and Reservations to the Convention on the Elimination of All Forms of Discrimination Against Women, available at http://www.unhchr.ch/html/menu3/b/treaty9_asp.htm.

73. See Human Rights Watch, "Perpetual Minors: Human Rights Abuses Stemming from Male Guardianship and Sex Segregation in Saudi Arabia," April 19, 2008, avail-able at http://www.hrw.org/en/reports/2008/04/19/perpetual-minors-0.

74. United Nations CEDAW/C/SAU/2 Convention on the Elimination of All Forms of Discrimination Against Women Distr.: General 29 March 2007; English Original: Arabic 07–29667 (E) 120507 230507*0729667* Committee on the Elimination of Discrimination against Women. Consideration of reports submitted by States Parties under article 18 of the Convention on the Elimination of All Forms of Discrimination against Women. Combined initial and second periodic reports of States Parties. Saudi Arabia, available at http://daccess-dds-ny.un.org/doc/UNDOC/GEN/N07/296/67/PDF/N0729667.pdf?OpenElement.

75. The Global Gender Gap Report 2010, available at http://www3.weforum.org/docs/WEF_GenderGap_Report_2010.pdf.

76. Committee on the Elimination of Discrimination Against Women. Pre-session working group. Fortieth session 14 January–1 February 2008. Responses to the list of issues and questions contained in document number CEDAW/C/SAU/Q/2, A.H. 1428 (A.D. 2007), available at http://www2.ohchr.org/english/bodies/cedaw/docs/CEDAW.C.SAU.Q.2.Add.1.pdf.

77. Ibid.

78. Zinnia Shah, "Saudi Arabia: Battle to Overturn Ban on Women Driving Is First Step to Women's Full Integration into Society," *Women Living Under Muslim Laws*, July 25, 2011, available at http://www.wluml.org/node/7459.

79. Neil MacFarquhar, "Saudis Arrest Woman Leading Right-to-Drive Campaign," *The New York Times*, May 23, 2011, available at http://www.nytimes.com/2011/05/24/world/middleeast/24saudi.html.

80. Suo Moto No. 1/K of 2006, Pakistan Citizenship Act. 1951, In Re: Gender Equality, available at http://federalshariatcourt.gov.pk/Leading%20Judgements/Suo%20Moto%20No.1-K%20of%202006.pdf.

81. Ibid., par. 3.

82. Ibid., par. 28.

83. Ibid., par. 27.

84. See "Pakistan: Whether the Spouse of a Citizen of Pakistan Can Acquire Citizenship; If So, Information Requirements, Procedures and Documents Needed," *Refworld*, October 28, 2010, available at http://www.unhcr.org/refworld/country,,IRBC,,PAK,,4dd1028e2,0.html.

85. Amnesty International, *Women in Afghanistan: A Human Rights Catastrophe*, AI Index: ASA 11/03/95.

86. See, generally, Nancy Hatch Dupree, "Afghan Women Under the Taliban," in *Fundamentalism Reborn: Afghanistan and the Taliban*, ed. William Maley (New York: New York University Press, 1998), 145–166; and Marjon E. Ghasemi, "Islam, Inter-national Human Rights, and Women's Equality: Afghan Women Under Taliban Rule," *Southern California Review of Law and Women's Studies* 8 (Spring 1999): 445–467.

87. See, for example, John Burns, "Sex and the Afghani Woman: Islam's Straightjacket," *The New York Times*, August 29, 1997, A4; "Islamic Rule Weighs Heavily for Afghans," *The New York Times*, September 24, 1997, A6.

88. See Anne E. Brodsky, *With All Our Strength: The Revolutionary Association of the Women of Afghanistan* (New York: Routledge, 2003).

89. Benazeer Roshan, "The More Things Change, the More They Stay the Same: The Plight of Afghan Women Two Years After the Overthrow of the Taliban," *Berkeley Women's Law Journal* 19 (2004): 270–286.

90. "Afghanistan 'Most Dangerous Place for Women,'" *Al Jazeera*, June 15, 2011, available at http://english.aljazeera.net/news/asia/2011/06/201161582525243992.html.

91. Lila Abu-Lughod, "Do Muslim Women Really Need Saving? Anthropological Reflections on Cultural Relativism and Its Others," *American Anthropologist* 104 (2002): 787; Mary Ann Franks, "Obscene Undersides: Women and Evil Between the Taliban and the United States," *HYPATIA* 18 (2003): 135–155.

92. See, for example, Condoleezza Rice's televised comments on the US involvement in the constitution drafting and her claim that "the United States stands for equality for women worldwide," adding: "We've communicated that very clearly to the Iraqi government," available at http://www.pbs.org/newshour/bb/white_house/july-dec05/rice_7-28.html.

93. See the discussions in Mayer, "Rhetorical Strategies," 104–114, and "Internationalization of the Conversation on Women's Rights: Arab Governments Face the CEDAW Committee," in *Islamic Law and the Challenge of Modernity*, eds. Yvonne Haddad and Barbara Freyer Stowasser (Walnut Creek, Calif.: Altamira Press, 2004), 147–154.

94. Tabandeh, *A Muslim Commentary*, 39.

95. Ibid., 41.

96. Ibid., 51.

97. Ibid., 52.

98. Mawdudi, *Purdah and the Status of Women in Islam*, 113–122.

99. Ibid., 120.

100. Ibid., 121–122.

101. Javad Bahonar, "Islam and Women's Rights," *al-Tawhid* 1 (1984): 160.

102. Ibid., 161.

103. Ibid., 165.

104. Ibid., 161.

105. Ibid., 164. Of course, since he was trying to show the Islamic treatment of women in a positive light, Bahonar had reason to avoid mentioning the discriminatory features of the inheritance scheme.

106. Hammed Shahidian, "Contesting Discourses of Sexuality in Post-Revolutionary Iran," in *Deconstructing Sexuality in the Middle East: Challenges and Discourses*, ed. Pinar Ilkkaracan (Burlington, VT: Ashgate, 2008), 117.

107. See Ann Elizabeth Mayer, "Islam and Human Rights: Different Issues, Different Contexts. Lessons from Comparisons," in *Islamic Law Reform and Human Rights: Challenges and Rejoinders*, eds. Tore Lindholm and Kari Vogt (Oslo: Nordic Human Rights Publications, 1993), 121–125.

108. Mary Daly, *The Church and the Second Sex* (Boston: Beacon Press, 1985), 85.
109. Ibid., 88.
110. Ibid., 154.
111. Ibid., 115.
112. Ibid.
113. Ibid., 87.

DISCUSSION QUESTIONS

1. According to the article, where does the violence against women in Pakistan stem from? Does it come from Islam? What are your thoughts on the issue of differentiating religion and culture?
2. Can a man who marries a Pakistani woman get Pakistani citizenship? Why is this allowed or not allowed? What are the implications of this law?
3. What country was named as more dangerous than the Congo or Somalia for women?
4. Describe the Iraqi constitution and how it limits the rights of women.
5. What did Ayatollah Javed Bahonar say about the characteristics of women in comparison to those of men?
6. What was said about the Catholic Church and sex stereotyping?

CONCLUSION

The post-modern age is full of possibility. It can also be so full of subjectivity and relativism that we risk losing all meaning. It is important to see the shifts of black and white to grey, but also to view these shifts with caution. Embracing grey should not mean that there is no longer a space for black and white. I believe the study of religion, like the study of any other subject, encompasses a full spectrum. Being able to navigate that spectrum is another issue entirely. The articles that make up this anthology have covered a wide range of topics: issues so heavily influenced by and laden with other issues that we can easily lose sight of what it all means. The discussion questions should have been a guide, primarily to help you observe and then describe what you were reading. This is an important first step when you are thinking about any issue regarding religion. Do not be discouraged to go back and read the articles a second or third time. Use a dictionary if you need to. Knowing the characteristics and appropriate vocabulary for an issue is essential. From there, these are the black and white concepts from which we can all start a dialogue.

Can we get to an understanding of the "grey area"? Surely, that understanding cannot be achieved from this introductory anthology alone. This is merely a stepping stone—one that can propel you to analyze, interpret, evaluate, or apply religion in a variety of ways. This is important because so much of our world is steeped in religion. It's almost impossible to turn on a news channel or your Facebook newsfeed without encountering information that in some way mentions religion. Ways of distinguishing what religion is and what role it plays in many of these news stories is a challenging task. Very often, religion gets lumped into a wide reaction of praise or blame. But should religion as a sole entity be praised or blamed unequivocally? This anthology has demonstrated that religions are simply concepts. We humans use these concepts to mold and shape our individual, communal, institutional, and political practices. Then, these practices inform our relationships and laws, as well as our treatment of the natural world, of others, and, most disparagingly, of women. Religious concepts are also not the sole influencing factor in that practice-molding process. This means that many blanket statements about religion deserve deeper analysis.

For scholars it is easy to see religion and its influencing factors, as well as its uses and limits. After years of study, being able to see these things easily might have become second nature for scholars. For students in introductory religious studies courses, however, this ability might not be as easily available. Still, this anthology should have shown the importance and relevance of studying religion, even if to a minor extent. Although there are about one billion agnostics and atheists in the world today, they do not live in a vacuum. Religion was a driving force behind the development of most civilizations. Today its rich history is still unfolding. Religion has been a powerful tool for thousands of years, and divorcing cultures from their religious upbringing is impossible. The next time you want to be dismissive of religion or allow someone else to dismiss it, remember just how much of the world was shaped, and is still being shaped, by this phenomenon we call religion.

CREDITS

1. John Grim and Mary Evelyn Tucker, Excerpt from: "Hinduism and the Transforming Affect of Devotion," *Ecology and Religion*, pp. 144–151, 241–244. Copyright © 2013 by Island Press. Reprinted with permission.
2. Sahana Udupa, Excerpt from: "Internet Hindus: Right-Wingers as New India's Ideological War," *Handbook of Religion and the Asian City: Aspiration and Urbanization in the Twenty-First Century*, ed. Peter van der Veer, pp. 434–443. Copyright © 2015 by University of California Press. Reprinted with permission.
3. Doranne Jacobson, Excerpt from: "Marriage: Women in India," *The Life of Hinduism*, ed. John Stratton Hawley and Vasudha Narayanan, pp. 67–71. Copyright © 2006 by University of California Press. Reprinted with permission.
4. David L. Gosling, Excerpt from: "Thailand: A Case Study," *Religion and Ecology in India and Southeast Asia*, pp. 92–100, 195–197. Copyright © 2001 by Taylor & Francis Group. Reprinted with permission.
5. Mahinda Deegalle, Excerpt from: "Jhu Politics for Peace and a Righteous State?," *Buddhism, Conflict and Violence in Modern Sri Lanka*, pp. 243–250. Copyright © 2006 by Taylor & Francis Group. Reprinted with permission.
6. Karma Lekshe Tsomo, Excerpt from: "Buddhist Nuns: Changes and Challenges," *Westward Dharma: Buddhism beyond Asia*, ed. Charles S. Prebish and Martin Baumann, pp. 260–271. Copyright © 2002 by University of California Press. Reprinted with permission.
7. Brianne Donaldson, Excerpt from: "'They'll Know We Are Process Thinkers by Our ...': *Finding the Ecological Ethic of Whitehead Through the Lens of Jainism and Ecofeminist Care*," Polydoxy: Theology of Multiplicity and Relation, ed. Catherine

Keller and Laurel Schneider, pp. 209–216. Copyright © 2010 by Taylor & Francis Group. Reprinted with permission.

8. Kim Skoog, Excerpt from: "The Morality of Sallekhaná: The Jaina Practice of Fasting to Death," *Jainism and Early Buddhism: Essays in Honor of Padmanabh S. Jaini*, ed. Olle Qvarnstrom, pp. 293–302. Copyright © 2003 by Jain Publishing Company. Reprinted with permission.

9. M. Whitney Kelting, Excerpt from: "Constructions of Femaleness in Jain Devotional Literature," *Jainism and Early Buddhism: Essays in Honor of Padmanabh S. Jaini*, ed. Olle Qvarnstrom, pp. 237–243. Copyright © 2003 by Jain Publishing Company. Reprinted with permission.

10. Hava Tirosh-Samuelson, Excerpt from: "Judaism and the Science of Ecology," *The Routledge Companion to Religion and Science*, ed. James W. Haag, Gregory R. Peterson, and Michael L. Spezio, pp. 346–355. Copyright © 2012 by Taylor & Francis Group. Reprinted with permission.

11. Robert D. Lee, Excerpt from: "The Transformation of Judaism in Israel," *Politics and Religion in the Middle East*, pp. 89–99, 302–305. Copyright © 2013 by Perseus Books Group. Reprinted with permission.

12. Karla Goldman, "A Worthier Place: Women, Reform Judaism, and the Presidents of the Hebrew Union College," *Contemporary Debates in American Reform Judaism: Conflicting Visions*, ed. Dana Evan Kaplan, pp. 171–179. Copyright © 2001 by Taylor & Francis Group. Reprinted with permission.

13. Paul Maltby, Excerpt from: "Fundamentalist Dominion, Postmodern Ecology," *Christian Fundamentalism and the Culture of Disenchantment*, pp. 113–120, 195–198. Copyright © 2013 by University of Virginia Press. Reprinted with permission.

14. Mark A. Smith, Excerpt from: "Homosexuality," *Secular Faith: How Culture Has Trumped Religion in American Politics*, pp. 128–137, 241–249. Copyright © 2015 by University of Chicago Press. Reprinted with permission.

15. Kelsy Burke, Excerpt from: "Sexual Awakening: Defining Women's Pleasure," *Christians under Covers: Evangelicals and Sexual Pleasure on the Internet*, pp. 108–116, 197–198. Copyright © 2016 by University of California Press. Reprinted with permission.

16. Ibrahim Abdul-Matin, "Green Muslims," *Green Deen: What Islam Teaches about Protecting the Planet*, pp. 46–56, 195–196. Copyright © 2010 by Berrett-Koehler Publishers. Reprinted with permission.

17. Robert A. Pape and James K. Feldman, Excerpt from: "Israel and Palestine," *Cutting the Fuse: The Explosion of Global Suicide Terrorism and How to Stop It*, pp. 232–240. Copyright © 2010 by University of Chicago Press. Reprinted with permission.
18. Ann Elizabeth Mayer, Excerpt from: "Restrictions on the Rights of Women," *Islam and Human Rights: Tradition and Politics*, pp. 124–131, 264–271. Copyright © 2012 by Perseus Books Group. Reprinted with permission.

Printed by Libri Plureos GmbH in Hamburg,
Germany